# GUIDED NOTEBOOK

## TO ACCOMPANY
## TRIGONOMETRY
### THIRD EDITION

### Kirk Trigsted
*University of Idaho*

The author and publisher of this book have used their best efforts in preparing this book. These efforts include the development, research, and testing of the theories and programs to determine their effectiveness. The author and publisher make no warranty of any kind, expressed or implied, with regard to these programs or the documentation contained in this book. The author and publisher shall not be liable in any event for incidental or consequential damages in connection with, or arising out of, the furnishing, performance, or use of these programs.

Reproduced by Pearson from electronic files supplied by the author.

Copyright © 2019 by Pearson Education, Inc. or its affiliates. All Rights Reserved. Printed in the United States of America. This publication is protected by copyright, and permission should be obtained from the publisher prior to any prohibited reproduction, storage in a retrieval system, or transmission in any form or by any means, electronic, mechanical, photocopying, recording, or otherwise. For information regarding permissions, request forms and the appropriate contacts within the Pearson Education Global Rights and Permissions Department, please visit www.pearsoned.com/permissions.

1 18

ISBN-13: 978-0-13-476811-3
ISBN-10: 0-13-476811-6

# Contents

| | | |
|---|---|---|
| **Chapter 1** | **An Introduction to Trigonometric Functions** | **1** |
| 1.1 | An Introduction to Angles: Degree and Radian Measure | 1 |
| 1.2 | Applications of Radian Measure | 13 |
| 1.3 | Triangles | 19 |
| 1.4 | Right Triangle Trigonometry | 27 |
| 1.5 | Trigonometric Functions of General Angles | 35 |
| 1.6 | The Unit Circle | 53 |
| **Chapter 2** | **The Graphs of Trigonometric Functions** | **63** |
| 2.1 | The Graphs of Sine and Cosine | 63 |
| 2.2 | More on Graphs of Sine and Cosine; Phase Shift | 83 |
| 2.3 | The Graphs of the Tangent, Cotangent, Secant, and Cosecant Functions | 101 |
| 2.4 | Inverse Trigonometric Functions I | 125 |
| 2.5 | Inverse Trigonometric Functions II | 137 |
| **Chapter 3** | **Trigonometric Identities, Formulas, and Equations** | **147** |
| 3.1 | Trigonometric Identities | 147 |
| 3.2 | The Sum and Difference Formulas | 159 |
| 3.3 | The Double-Angle and Half-Angle Formulas | 169 |
| 3.4 | The Product-to-Sum and Sum-to-Product Formulas | 181 |
| 3.5 | Trigonometric Equations | 189 |
| **Chapter 4** | **Applications of Trigonometry** | **203** |
| 4.1 | Right Triangle Applications | 203 |
| 4.2 | The Law of Sines | 211 |
| 4.3 | The Law of Cosines | 223 |
| 4.4 | Area of Triangles | 231 |
| **Chapter 5** | **Polar Equations, Complex Numbers, and Vectors** | **237** |
| 5.1 | Polar Coordinates and Polar Equations | 237 |
| 5.2 | Graphing Polar Equations | 249 |
| 5.3 | Complex Numbers in Polar Form; DeMoivre's Theorem | 275 |
| 5.4 | Vectors | 291 |
| 5.5 | The Dot Product | 307 |

# Section 1.1 Guided Notebook

## 1.1 An Introduction to Angles: Degree and Radian Measure
- ☐ Work through Section 1.1 TTK #1
- ☐ Work through Section 1.1 Objective 1
- ☐ Work through Section 1.1 Objective 2
- ☐ Work through Section 1.1 Objective 3
- ☐ Work through Section 1.1 Objective 4
- ☐ Work through Section 1.1 Objective 5

## Section 1.1 An Introduction to Angles: Degree and Radian Measure

### 1.1 Things To Know

1. Sketching the Graph of a Circle

Can you sketch the graph of a circle? Try working through a "You Try It" problem or watch the video.

### Section 1.1 Introduction

What is the definition of a **vertex**?

What is the definition of the **initial side**?

What is the definition of the **terminal side**?

Sketch an angle with positive measure, labeling the vertex, initial side, and terminal side. Do the same for an angle with negative measure.

Section 1.1

What does it mean for an angle to be in **standard position**?

Sketch a coordinate plane and label the **four quadrants**.

Section 1.1  Objective 1 Understanding Degree Measure

In the **degree measure** system, what is the symbol used to indicate a degree? How many degrees are in a one complete counterclockwise rotation?

Sketch three coordinate planes, illustrating angles of 360, 90, and –45 degrees, respectively. (See Figures 3, 4, and 5.)

What is the definition of an **acute angle**?

What is the definition of an **obtuse angle**?

What is the definition of a **quadrantal angle**?

What is the term for an angle of exactly 90 degrees?

What is the term for an angle of exactly 180 degrees?

Section 1.1

- What does it mean for angles to be **coterminal**?

Sketch the two coordinate planes illustrating common positive and negative angles as seen in Figure 6.

- Work through the video accompanying Example 1 showing all work below. Draw each angle in standard position and state the quadrant in which the terminal side of the angle lies or the axis on which the terminal side of the angle lies.

  a. $\theta = 60°$

  b. $\alpha = -270°$

  c. $\beta = 420°$

3

Copyright © 2019 Pearson Education, Inc.

Section 1.1

## Section 1.1 Objective 2 Finding Coterminal Angles Using Degree Measure

What is the definition of **coterminal angles**?

Starting with a given angle, how can you obtain coterminal angles? (See the **coterminal angle** definition box.)

What notation is used to denote the angle of least nonnegative measure that is coterminal with $\theta$?

Work through the video with Example 2 and show all work below.
Find the angle of least nonnegative measure, $\theta_C$, that is coterminal with $\theta = -697°$.

Section 1.1 Objective 3 Understanding Radian Measure

Carefully work through the **animation** seen next to Objective 3 on page 1.1-13 and answer the questions below:

Draw a circle centered at the origin having a radius of $r$ units. What is the equation of the circle?

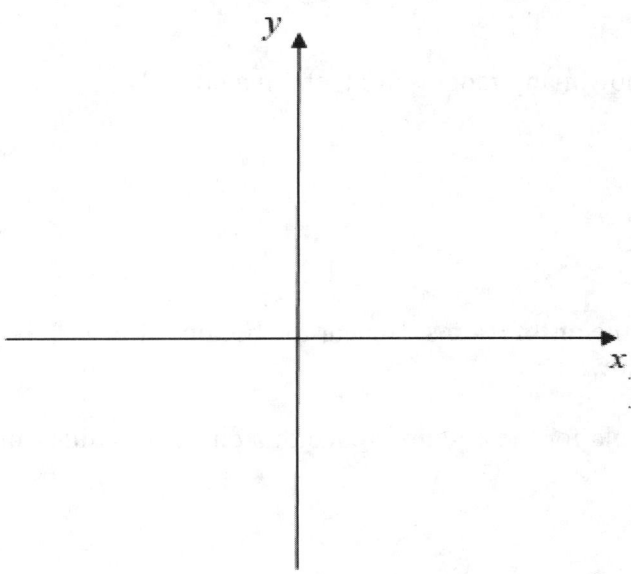

What is the definition of a **central angle**?

What is the definition of an **intercepted arc**? What variable is typically used to represent it?

On the graph of your circle, above, draw a central angle so that the intercepted arc is the same length as the radius of your circle.

What is the measure of this central angle called?

Section 1.1

What is the definition of a **radian**?

Approximately how many radians are there in a circle?

Carefully work through the **animation** seen near the bottom of page 1.1-13 and answer the questions below:

What is the formula for the circumference of a circle of radius *r* units?

(Fill in the blank) A central angle of 360° intercepts an arc length of _____.

Complete the proportion below as seen in the animation.

$$\frac{360°}{\Box} = \frac{\Box}{r}$$

Fill in the box: $360° = \Box$ radians.
Fill in the box: $180° = \Box$ radians.

6

Sketch three coordinate planes, illustrating angles of $2\pi$, $\dfrac{\pi}{2}$, and $\dfrac{-\pi}{4}$ radians, respectively. (See Figures 10, 11, and 12.)

Sketch two coordinate planes illustrating common positive and negative angles in radians as seen in Figure 13.

Section 1.1

Work through the interactive video accompanying Example 3, showing all work below. Draw each angle in standard position and state the quadrant in which the terminal side of the angle lies or the axis on which the terminal side of the angle lies.

a. $\theta = \dfrac{\pi}{3}$

b. $\alpha = -\dfrac{3\pi}{2}$

c. $\beta = \dfrac{7\pi}{3}$

Section 1.1

Section 1.1 Objective 4 Converting between Degree Measure and Radian Measure

To convert **degrees to radians**, multiply by _____.

To convert **radians to degrees**, multiply by _____.

Work through the interactive video with Example 4 and show all work below.
Convert each angle given in degree measure into radians.

a. 45°

b. −150°

c. 56°

Section 1.1

Work through the interactive video with Example 5 and show all work below. Convert each angle given in radian measure into degrees. Round to two decimal places if needed.

a. $\dfrac{2\pi}{3}$ radians

b. $-\dfrac{11\pi}{6}$ radians

c. 3 radians

Section 1.1  Objective 5 Finding Coterminal Angles Using Radian Measure

For any angle $\theta$ and for any nonzero integer $k$, we can find a coterminal angle using what expression?

Work through Example 6 and show all work below.

Find three angles that are coterminal with $\theta = \dfrac{\pi}{3}$ using $k=1$, $k=-1$, and $k=2$.

Section 1.1

Work through the video with Example 7 and show all work below.

Find the angle of least nonnegative measure, $\theta_C$, that is coterminal with $\theta = -\dfrac{21\pi}{4}$.

## Section 1.2 Guided Notebook

### 1.2 Applications of Radian Measure
- ☐ Work through Section 1.2 TTK #1
- ☐ Work through Section 1.2 Objective 1
- ☐ Work through Section 1.2 Objective 2
- ☐ Work through Section 1.2 Objective 3

## Section 1.2 Applications of Radian Measure

### 1.2 Things To Know

1. Converting between Degree Measure and Radian Measure

Try working through a "You Try It" problem or watch the animation and/or interactive video.

Section 1.2 Objective 1 Determining the Area of a Sector of a Circle

What is the definition of a **sector** of a circle?

What is the formula for the **area of a sector of a circle**?

Work through Example 1 showing all work below.
Find the area of the sector of a circle of radius 15 inches formed by a central angle of $\theta = \frac{\pi}{10}$ radians. Round the answer to two decimal places.

Section 1.2

Use the Caution note that follows Example 1 to fill in the following blanks:

The formula used to find the area of a sector of a circle is only valid if the angle is given in
_____.

An angle given in _____ must first be converted to _____.


Work through the video with Example 2 showing all work below.
Find the area of the sector of a circle of diameter 21 meters formed by a central angle of $135°$. Round the answer to two decimal places.




Section 1.2 Objective 2 Computing the Arc Length of a Sector of a Circle

What is the relationship between the central angle on a circle of radius $r$ and the length of the intercepted arc, $s$? (Hint: See the **Arc Length of a Sector of a Circle** text box.)




Work through Example 3 and show all work below.

Find the length of the arc intercepted by a central angle of $\theta = \dfrac{\pi}{6}$ in a circle of radius $r = 22$ centimeters. Round the answer to two decimal places.

Section 1.2

Work through the video with Example 4 and show all work below.
A 120° central angle intercepts an arc of 23.4 inches. Calculate the radius of the circle. Round the answer to two decimal places.

Work through the video with Example 5 and show all work below.
Two gears are connected so that the smaller gear turns the larger gear. When the smaller gear with a radius of 3.6 centimeters rotates 300°, how many degrees will the larger gear with a radius of 7.5 cm rotate?

Section 1.2

## Section 1.2 Objective 3 Understanding Angular Velocity and Linear Velocity

What is the definition of **angular velocity**?



What is the definition of **linear velocity**?



Work through the video that accompanies Example 6 showing all work below.
Determine the angular velocity and the linear velocity of a point on the equator of the Earth. Assume that the radius of the Earth at the equator is 3963 miles.

Work through Example 7 showing all work below.
The propeller of a small airplane is rotating 1500 revolutions per minute.
Find the angular velocity of the propeller in radians per minute.

Work through Example 8 showing all work below.
In 2008, the old Ferris wheel at the world-famous Santa Monica pier in Santa Monica, California, was replaced with a new solar-powered Ferris wheel that contains thousands of energy-efficient LED (light-emitting diode) lights that can illuminate into many different colors and designs. The wheel is approximately 85 feet in diameter and rotates 2.5 revolutions per minute. Find the linear velocity (in mph) of this new Ferris wheel at the outer edge. Round to two decimal places.

Section 1.2

What is the **Relationship between Linear Velocity and Angular Velocity**?

Work through Example 9 showing all work below.
Suppose that the propeller blade from Example 7 is 2.5 meters in diameter. Find the linear velocity in meters per minute for a point located on the tip of the propeller.

## Section 1.3 Guided Notebook

### 1.3 Triangles
- [ ] Work through Section 1.3 TTK #1
- [ ] Work through Section 1.3 Objective 1
- [ ] Work through Section 1.3 Objective 2
- [ ] Work through Section 1.3 Objective 3
- [ ] Work through Section 1.3 Objective 4
- [ ] Work through Section 1.3 Objective 5

## Section 1.3 Triangles

### 1.3 Things To Know

1. Converting between Degree Measure and Radian Measure

Try working through a "You Try It" problem or watch the animation and/or interactive video.

Section 1.3 Objective 1 Classifying Triangles

What does it mean for two angles or sides of a triangle to be **congruent**?

What is an **acute triangle**?

What is an **obtuse triangle**?

What is a **right triangle**?

Section 1.3

Sketch and label an acute, obtuse, and right triangle, as seen in Figure 18.




What is a **scalene triangle**?


What is an **isosceles triangle**?


What is an **equilateral triangle**?


Sketch a scalene, isosceles, and equilateral triangle, as seen in Figure 19.




Work through Example 1 showing all work below.
Classify the given triangle as acute, obtuse, right, scalene, isosceles, or equilateral. State all that apply.

20

Copyright © 2019 Pearson Education, Inc.

Section 1.3

Section 1.3 Objective 2 Using the Pythagorean Theorem

What is the **Pythagorean Theorem**?

Work through the video that accompanies Example 2 and show all work below. Use the Pythagorean Theorem to find the length of the missing side of the given right triangle.

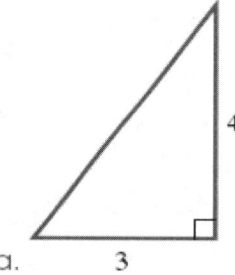
a.   3

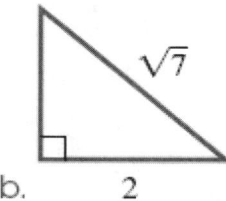
b.   2

21

Section 1.3

Work through the video with Example 3 and show all work below.
A Major League baseball diamond is really a square. The distance between each consecutive base is 90 feet. What is the distance between home plate and second base? Round to two decimal places.

Section 1.3 Objective 3 Understanding Similar Triangles

What is the definition of **similar triangles**?

What are the **Properties of Similar Triangles**?

1.

2.

Work through the video accompanying Example 4 showing all work below.
Triangles $ABC$ and $XYZ$ are similar. Find the lengths of the missing sides of triangle $ABC$.

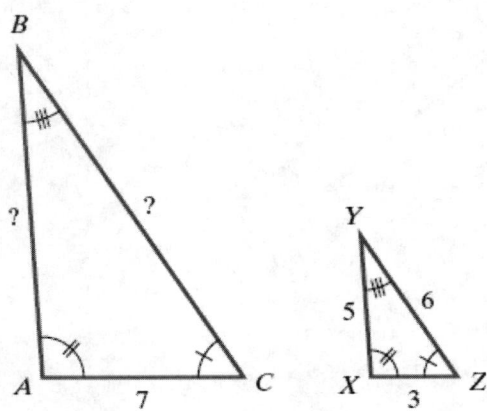

Section 1.3

- What is the definition of the **Proportionality Constant of Similar Triangles**?

Work through the animation accompanying Example 5 showing all work below. The triangles below are similar. Find the proportionality constant. Then find the lengths of the missing sides.

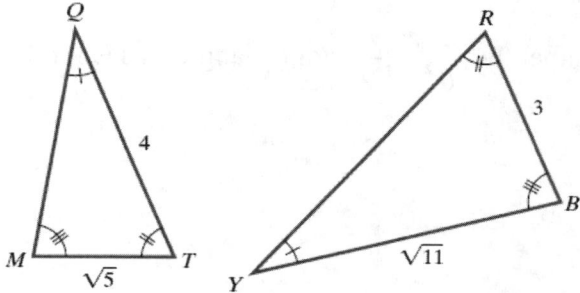

- 

Work through the video accompanying Example 6 showing all work below. The right triangles below are similar. Determine the lengths of the missing sides.

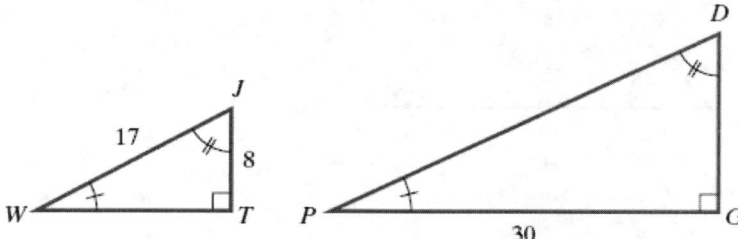

- 

23

Copyright © 2019 Pearson Education, Inc.

Section 1.3

## Section 1.3 Objective 4 Understanding the Special Right Triangles

Sketch and label the $\frac{\pi}{4}, \frac{\pi}{4}, \frac{\pi}{2}$ right triangle as seen in Figure 23.

Sketch and label the $\frac{\pi}{6}, \frac{\pi}{3}, \frac{\pi}{2}$ right triangle as seen in Figure 27.

Work through the interactive video with Example 7 and show all work below.
Determine the lengths of the missing sides of each right triangle.

a.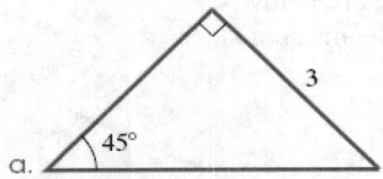

b.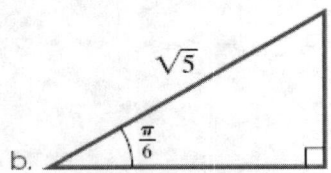

## Section 1.3 Objective 5 Using Similar Triangles to Solve Applied Problems

Work through Example 8 and show all work below.
The shadow of a cell tower is 80 feet long. A boy 3 feet 9 inches tall is standing next to the tower. If the boy's shadow is 6 feet long, find the height of the cell tower.

Work through the video with Example 9 and show all work below.
Two people are standing on opposite sides of a small river. One person is located at point $Q$, a distance of 20 feet from a bridge. The other person is standing on the southeast corner of the bridge at point $P$. The angle between the bridge and the line of sight from $P$ to $Q$ is $30°$. Use this information to determine the length of the bridge and the distance between the two people. Round your answer to two decimal places as needed.

## Section 1.4 Guided Notebook

### 1.4 Right Triangle Trigonometry
- ☐ Work through Section 1.4 TTK #1–3
- ☐ Work through Section 1.4 Objective 1
- ☐ Work through Section 1.4 Objective 2
- ☐ Work through Section 1.4 Objective 3
- ☐ Work through Section 1.4 Objective 4
- ☐ Work through Section 1.4 Objective 5

## Section 1.4 Right Triangle Trigonometry

### 1.4 Things To Know

1. Converting between Degree Measure and Radian Measure

Try working through a "You Try It" problem or watch the animation and/or interactive video.

2. Understanding Similar Triangles

Try working through a "You Try It" problem or watch the video.

3. Understanding the Special Right Triangles

Try working through a "You Try It" problem or watch the animation.

Section 1.4

<u>Section 1.4 Objective 1 Understanding the Right Triangle Definitions of the Trigonometric Functions</u>

Sketch and label the two triangles seen in Figure 29.

What are the **Right Triangle Definitions of the Trigonometric Functions**?

Work through the interactive video with Example 1 showing all work below.
Given the right triangle evaluate the six trigonometric functions of the acute angle $\theta$.

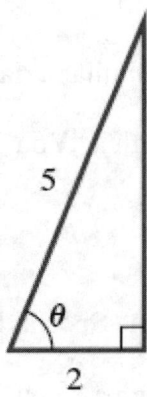

Section 1.4

Work through the video with Example 2 showing all work below.
If $\theta$ is an acute angle of a right triangle and if $\sin\theta = \frac{3}{4}$, then find the values of the remaining five trigonometric functions for angle $\theta$.

Section 1.4 Objective 2 Using the Special Right Triangles

Copy down the trigonometric functions for acute angles $\frac{\pi}{6}, \frac{\pi}{4}, \frac{\pi}{3}$ as seen in Table 1.

| $\theta$ | $\frac{\pi}{6}(30°)$ | $\frac{\pi}{4}(45°)$ | $\frac{\pi}{3}(60°)$ |
|---|---|---|---|
| $\sin\theta$ | | | |
| $\cos\theta$ | | | |
| $\tan\theta$ | | | |

Section 1.4

Work through the video with Example 3 and show all work below.
Determine the value of $\csc\dfrac{\pi}{6}+\cot\dfrac{\pi}{4}$.

Work through Example 4 and show all work below.
Determine the measure of the acute angle $\theta$ for which $\sec\theta = 2$.

Section 1.4

Section 1.4 Objective 3 Understanding the Fundamental Trigonometric Identities

What are the **Quotient Identities**?

What are the **Reciprocal Identities**?

Work through the video that accompanies Example 5 showing all work below.
Given that $\sin\theta = \dfrac{5}{7}$ and $\cos\theta = \dfrac{2\sqrt{6}}{7}$, find the values of the remaining four trigonometric functions using identities.

Section 1.4

What are the **Pythagorean Identities**?

Work through the interactive video with Example 6 showing all work below.
Use identities to find the exact value of each trigonometric expression.

a. $\tan 37° - \dfrac{\sin 37°}{\cos 37°}$

b. $\dfrac{1}{\cos^2 \dfrac{\pi}{9}} - \dfrac{1}{\cot^2 \dfrac{\pi}{9}}$

Section 1.4  Objective 4 Understanding Cofunctions

What does it mean for two angles to be **complementary**?

Section 1.4

What are the **Cofunction Identities**?

Work through the interactive video with Example 7 and show all work below.

a. Rewrite the expression $(\cot(\frac{\pi}{2}-\theta))\cos\theta$ as one of the six trigonometric functions of acute angle $\theta$.

b. Determine the exact value of $\sec 55° \csc 35° - \tan 55° \cot 35°$.

Section 1.4

Section 1.4 Objective 5 Evaluating Trigonometric Functions Using a Calculator

Work through the video with Example 8 and show all work below.
Evaluate each trigonometric expression using a calculator. Round each answer to four decimal places.

a. $\sin \dfrac{8\pi}{7}$

b. $\cot 70°$

c. $\sec \dfrac{\pi}{5}$

Write down the **CAUTION** as seen on page 1.4-29:

## Section 1.5 Guided Notebook

### 1.5 Trigonometric Functions of General Angles
- [ ] Work through Section 1.5 TTK #1
- [ ] Work through Section 1.5 TTK #2
- [ ] Work through Section 1.5 TTK #3
- [ ] Work through Section 1.5 TTK #4
- [ ] Work through Section 1.5 TTK #5
- [ ] Work through Section 1.5 Objective 1
- [ ] Work through Section 1.5 Objective 2
- [ ] Work through Section 1.5 Objective 3
- [ ] Work through Section 1.5 Objective 4
- [ ] Work through Section 1.5 Objective 5
- [ ] Work through Section 1.5 Objective 6

### Section 1.5 Trigonometric Functions of General Angles

#### 1.5 Things To Know

1. Understanding Radian Measure
Try working through a "You Try It" problem or watch the interactive video.

2. Converting between Degree Measure and Radian Measure
Try working through a "You Try It" problem or watch the animation and/or interactive video.

3. Finding Coterminal Angles Using Radian Measure
Try working through a "You Try It" problem.

Section 1.5

4. Understanding the Special Right Triangle
Try working through a "You Try It" problem or watch the animation.

5. Understanding the Right Triangle Definitions of the Trigonometric Functions
Try working through a "You Try It" problem or watch the video.

Section 1.5  Objective 1 Understanding the Four Families of Special Angles

What is the **Quadrantal Family of Angles**? Sketch the angles shown in Figure 35.

What is the $\frac{\pi}{3}$ **Family of Angles**? Sketch the angles shown in Figure 36.

Section 1.5

● What is the $\dfrac{\pi}{6}$ **Family of Angles?** Sketch the angles shown in Figure 37.

What is the $\dfrac{\pi}{4}$ **Family of Angles?** Sketch the angles shown in Figure 38.

● Work through the interactive video with Example 1 showing all work below.
Each of the given angles belongs to one of the four families of special angles. Determine the family of angles to which it belongs, sketch the angle, and then determine the angle of least nonnegative measure, $\theta_C$, coterminal with the given angle.

a. $\theta = \dfrac{29\pi}{6}$

b. $\theta = \dfrac{14\pi}{2}$

c. $\theta = -\dfrac{18\pi}{4}$

●

Section 1.5

d.  $\theta = \dfrac{11\pi}{4}$

e.  $\theta = \dfrac{14\pi}{6}$

f.  $\theta = 420°$

g.  $\theta = -495°$

Section 1.5

Section 1.5 Objective 2 Understanding the Definitions of the Trigonometric Functions of General Angles

What are the **General Angle Definitions of the Trigonometric Functions**?

Under what conditions will the following trigonometric functions be undefined (if ever)?

$\tan \theta = \dfrac{y}{x}$ and $\sec \theta = \dfrac{r}{x}$:

$\csc \theta = \dfrac{r}{y}$ and $\cot \theta = \dfrac{x}{y}$:

$\sin \theta = \dfrac{y}{r}$ and $\cos \theta = \dfrac{x}{r}$:

Section 1.5

Work through the video with Example 2 and show all work below.
Suppose that the point $(-4,-6)$ is on the terminal side of an angle $\theta$.
Find the six trigonometric functions of $\theta$.

**Section 1.5 Objective 3 Finding the Values of the Trigonometric Functions of Quadrantal Angles**

The value of the six trigonometric functions of a quadrantal angle is 0, 1, −1, or undefined. Fill in the table below. This table represents the values of the six trigonometric functions of quadtrantal angles.

| $\theta$ | $\sin\theta$ | $\cos\theta$ | $\tan\theta$ | $\csc\theta$ | $\sec\theta$ | $\cot\theta$ |
|---|---|---|---|---|---|---|
| $0$ | | | | | | |
| $\dfrac{\pi}{2}$ | | | | | | |
| $\pi$ | | | | | | |
| $\dfrac{3\pi}{2}$ | | | | | | |

Section 1.5

Work through the interactive video accompanying Example 3 showing all work below. Without using a calculator, determine the value of the trigonometric function or state that the value is undefined.

a. $\cos(-11\pi)$

b. $\csc(-270°)$

c. $\tan\left(\dfrac{13\pi}{2}\right)$

d. $\sin(540°)$

e. $\cot\left(-\dfrac{7\pi}{2}\right)$

Section 1.5

## Section 1.5  Objective 4 Understanding the Signs of the Trigonometric Functions

The sign of each trigonometric function is determined by the _____ in which the terminal side of the angle lies.

Which trigonometric functions are positive for all angles with a terminal side lying in the following quadrants?

    Quadrant I:

    Quadrant II:

    Quadrant III:

    Quadrant IV:

What acronym can help us remember the signs of the trigonometric functions for angles whose terminal side lies in one of the four quadrants?

Sketch the diagram shown in Figure 49.

Three grids are shown below. (See Figure 50.) The first grid represents the sign of the values of $y = \sin x$ in each quadrant. The middle grid represents the sign of the values of $y = \cos x$. The third quadrant represents the sign of the values of $y = \tan x$ in each quadrant. Place a "+" or "−" in each quadrant of each grid to represent the appropriate sign.

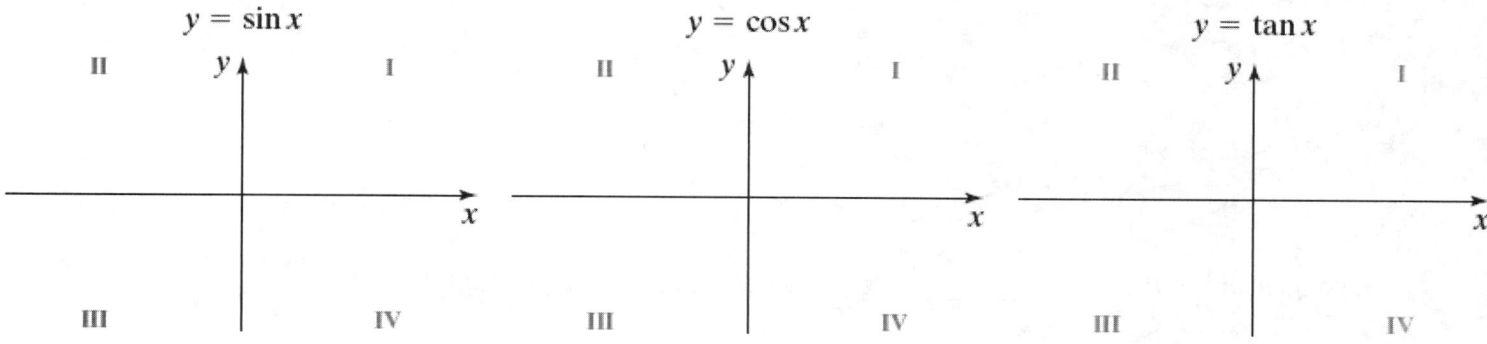

Work through the video with Example 4 and show all work below.
Suppose $\theta$ is a positive angle in standard position such that $\sin\theta < 0$ and $\sec\theta > 0$.

a. Determine the quadrant in which the terminal side of angle $\theta$ lies.

b. Find the value of $\tan\theta$ if $\sec\theta = \sqrt{5}$.

Section 1.5

## Section 1.5 Objective 5 Determining Reference Angles

What is the definition of the **Reference Angle**?

The measure of the _____ $\theta_R$ depends on the quadrant in which the _____ of $\theta_C$ lies.

The four cases for reference angles are shown below. Fill in the blanks.

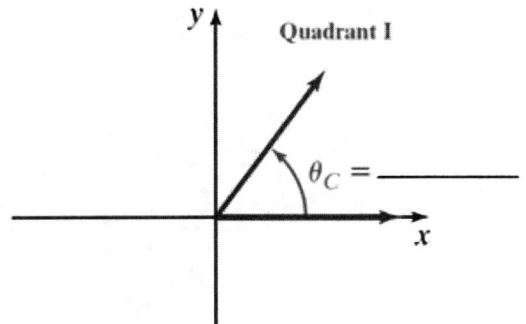

**Case 1:** If the terminal side of $\theta_C$ lies in Quadrant I, then $\theta_R =$ _____ .

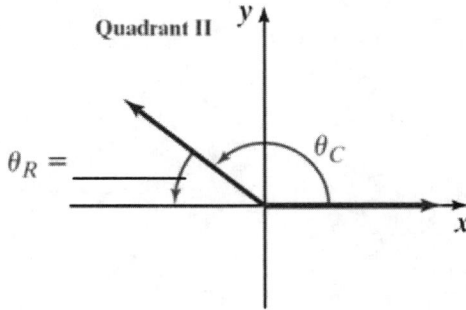

**Case 2:** If the terminal side of $\theta_C$ lies in Quadrant II, then $\theta_R =$ _____ (or $\theta_R =$ _____ ).

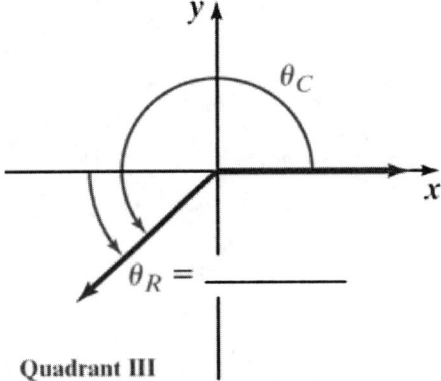

**Case 3:** If the terminal side of $\theta_C$ lies in Quadrant III, then $\theta_R =$ _____ (or $\theta_R =$ _____ ).

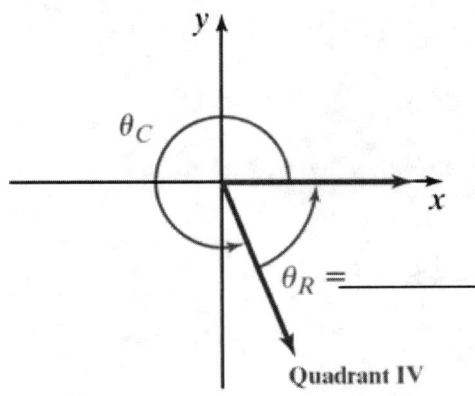

**Case 4:** If the terminal side of $\theta_C$ lies in Quadrant IV, then $\theta_R =$ _____ (or $\theta_R =$ _____ ).

Section 1.5

Work through the interactive video with Example 5 and show all work below.
For each of the given angles, determine the reference angle.

a. $\theta = \dfrac{5\pi}{3}$

b. $\theta = \dfrac{11\pi}{4}$

c. $\theta = -\dfrac{25\pi}{6}$

d. $\theta = \dfrac{16\pi}{6}$

Section 1.5

Work through the interactive video with Example 6 and show all work below.
For each of the given angles, determine the reference angle.

a. $\theta = \dfrac{5\pi}{8}$

b. $\theta = \dfrac{22\pi}{9}$

c. $\theta = -\dfrac{5\pi}{7}$

Section 1.5

Work through the interactive video with Example 7 and show all work below.
For each of the given angles, determine the reference angle.

a. $\theta = 225°$

b. $\theta = -233°$

c. $\theta = 510°$

Section 1.5

## Section 1.5 Objective 6 Evaluating Trigonometric Functions of Angles Belonging to the pi/3, pi/6, or pi/4 Families

Work through the interactive video with Example 8 and show all work below.

Find the values of the six trigonometric functions for $\theta = \dfrac{7\pi}{4}$.

What are the four **Steps for Evaluating Trigonometric Functions of Angles Belonging to the $\dfrac{\pi}{3}, \dfrac{\pi}{6},$ or $\dfrac{\pi}{4}$ Families?**

**Step 1:**

**Step 2:**

**Step 3:**

**Step 4:**

Section 1.5

Work through the interactive video with Example 9 and show all work below.
Find the exact value of each trigonometric expression without using a calculator.

a. $\sin\left(\dfrac{7\pi}{6}\right)$

b. $\cot\left(-\dfrac{22\pi}{3}\right)$

c. $\tan\left(\dfrac{11\pi}{4}\right)$

d. $\cos\left(\dfrac{11\pi}{3}\right)$

e. $\sec\left(\dfrac{5\pi}{6}\right)$

f. $\csc\left(-\dfrac{7\pi}{6}\right)$

## Section 1.6 Guided Notebook

**1.6 The Unit Circle**
- ☐ Work through Section 1.6 TTK #1–6
- ☐ Work through Section 1.6 Objective 1
- ☐ Work through Section 1.6 Objective 2
- ☐ Work through Section 1.6 Objective 3
- ☐ Work through Section 1.6 Objective 4
- ☐ Work through Section 1.6 Objective 5

## Section 1.6 The Unit Circle

### 1.6 Things To Know

1. Converting between Degree Measure and Radian Measure
Try working through a "You Try It" problem or watch the animation and/or interactive video.

2. Understanding the Special Right Triangle
Try working through a "You Try It" problem or watch the interactive video.

3. Understanding the Right Triangle Definitions of the Trigonometric Functions
Try working through a "You Try It" problem.

Section 1.6

4. Understanding the Four Families of Special Angles
Try working through a "You Try It" problem or watch the interactive video.

5. Understanding the Definitions of the Trigonometric Functions of General Angles
Try working through a "You Try It" problem or watch the video and/or animation.

6. Evaluating Trigonometric Functions of Angles Belonging to the $\frac{\pi}{3}$, $\frac{\pi}{6}$, or $\frac{\pi}{4}$ Families
Try working through a "You Try It" problem or refer to Section 6.5 or watch the interactive video.

Section 1.6

## Section 1.6 Objective 1 Understanding the Definition of the Unit Circle

What is the **Unit Circle**? (Define it in words and sketch it.)

Work through the interactive video with Example 1 showing all work below. Determine the missing coordinate of a point that lies on the graph of the unit circle given the quadrant in which the point is located.

a. $(-\frac{1}{8}, y)$; Quadrant III

b. $(x, -\frac{\sqrt{3}}{2})$; Quadrant IV

c. $(-\frac{1}{\sqrt{2}}, y)$; Quadrant II

Section 1.6

Section 1.6 Objective 2 Using Symmetry to Determine Points That Lie on the Unit Circle

What are the three ways in which the unit circle is symmetric?

Work through the video with Example 2 and show all work below.

Verify that the point $(-\frac{1}{8}, -\frac{3\sqrt{7}}{8})$ lies on the graph of the unit circle. Then use symmetry to find three other points that also lie on the graph of the circle.

Section 1.6 Objective 3 Understanding the Unit Circle Definitions of the Trigonometric Functions

The _____ of a sector of the unit circle is exactly equal

to the measure of the _____.

Section 1.6

- What are the **Unit Circle Definitions of the Trigonometric Functions**?
  (Include the six equations and sketch the graph.)

- Work through the video with Example 3 showing all work below.

  If $\left(-\dfrac{1}{4}, \dfrac{\sqrt{15}}{4}\right)$ is a point on the unit circle corresponding to a real number $t$, find the values of the six trigonometric functions of $t$.

## Section 1.6 Objective 4 Using the Unit Circle to Evaluate Trigonometric Functions at Increments of π/2

Use the unit circle definitions of the trigonometric functions and the figure below to fill out the table.

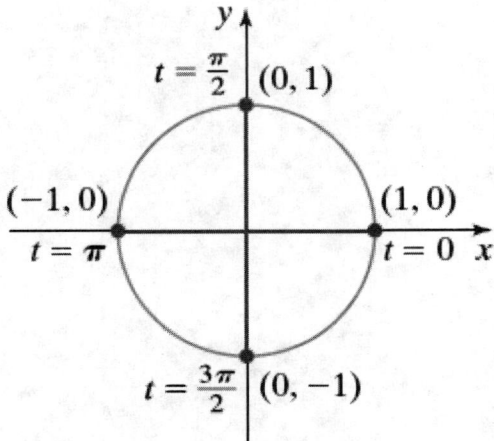

| $t$ | sin $t$ | cos $t$ | tan $t$ | csc $t$ | sec $t$ | cot $t$ |
|---|---|---|---|---|---|---|
| $0, 2\pi$ | | | | | | |
| $\dfrac{\pi}{2}$ | | | | | | |
| $\pi$ | | | | | | |
| $\dfrac{3\pi}{2}$ | | | | | | |

Section 1.6

Work through the interactive video with Example 4 and show all work below.
Use the unit circle to determine the value of each expression or state that it is undefined.

a. $\cos 11\pi$

b. $\tan \dfrac{5\pi}{2}$

c. $\csc\left(-\dfrac{3\pi}{2}\right)$

Section 1.6

## Section 1.6 Objective 5 Using the Unit Circle to Evaluate Trigonometric Functions for Increments of $\pi/6$, $\pi/4$, and $\pi/3$

Sketch and label the unit circle as seen in Figure 70.

Section 1.6

Work through the animation with Example 5 and show all work below.
Use the unit circle to determine the following values.

a. $\tan\left(\dfrac{7\pi}{3}\right)$

b. $\sin\left(-\dfrac{3\pi}{4}\right)$

c. $\sec(480°)$

d. $\csc\left(-\dfrac{13\pi}{3}\right)$

Section 1.6

Carefully work through the **Guided Visualization** seen on page 1.6-23.
Choose two values of $t$ (at least one negative). Sketch a unit circle, label the point that lies on the unit circle for each value of $t$, and find the values of the six trigonometric functions for each of your values of $t$. Below is an example:

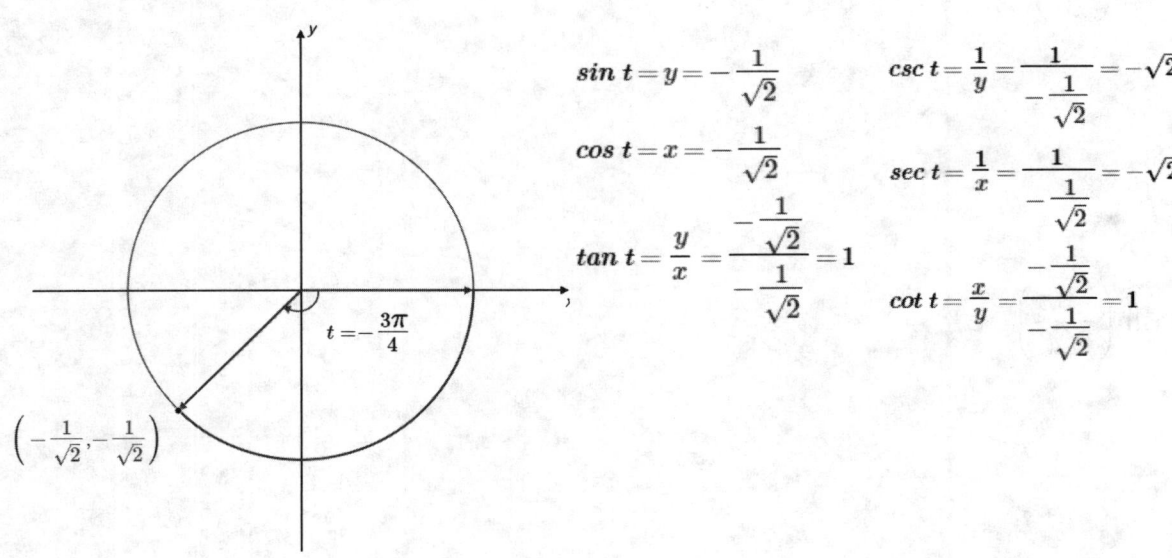

## Section 2.1 Guided Notebook

### 2.1 The Graphs of Sine and Cosine
- ☐ Work through Section 2.1 TTK #1–6
- ☐ Work through Section 2.1 Objective 1
- ☐ Work through Section 2.1 Objective 2
- ☐ Work through Section 2.1 Objective 3
- ☐ Work through Section 2.1 Objective 4
- ☐ Work through Section 2.1 Objective 5
- ☐ Work through Section 2.1 Objective 6

### Section 2.1 The Graphs of Sine and Cosine

### 2.1 Things To Know

1. Determining Whether a Function is Even, Odd, or Neither
Try working through a "You Try It" problem or watch the video.




2. Using Vertical Stretches and Compressions to Graph Functions
Try working through a "You Try It" problem or watch the video.




3. Using Horizontal Stretches and Compressions to Graph Functions
Try working through a "You Try It" problem or watch the video.




4. Using the Special Right Triangles
Try working through a "You Try It" problem or watch the video.

Section 2.1

5. Finding the Values of the Trigonometric Functions of Quadrantal Angles
Try working through a "You Try It" problem or watch the video.

6. Evaluating Trigonometric Functions of Angles Belonging to the $\frac{\pi}{3}$, $\frac{\pi}{6}$, or $\frac{\pi}{4}$ Families
Try working through a "You Try It" problem or watch the interactive video.

Section 2.1 Objective 1 Understanding the Graph of the Sine Function and Its Properties

Watch the video that accompanies Objective 1 and fill in the tables below.

| $x$ | 0 | $\frac{\pi}{2}$ | $\pi$ | $\frac{3\pi}{2}$ | $2\pi$ |
|---|---|---|---|---|---|
| $y = \sin x$ | | | | | |

| $x$ | $\frac{\pi}{6}$ | $\frac{5\pi}{6}$ | $\frac{7\pi}{6}$ | $\frac{11\pi}{6}$ |
|---|---|---|---|---|
| $y = \sin x$ | | | | |

| $x$ | $\frac{\pi}{4}$ | $\frac{3\pi}{4}$ | $\frac{5\pi}{4}$ | $\frac{7\pi}{4}$ |
|---|---|---|---|---|
| $y = \sin x$ | | | | |

| $x$ | $\frac{\pi}{3}$ | $\frac{2\pi}{3}$ | $\frac{4\pi}{3}$ | $\frac{5\pi}{3}$ |
|---|---|---|---|---|
| $y = \sin x$ | | | | |

Sketch the graph of $y = \sin x$ on the interval $[0, 2\pi]$.

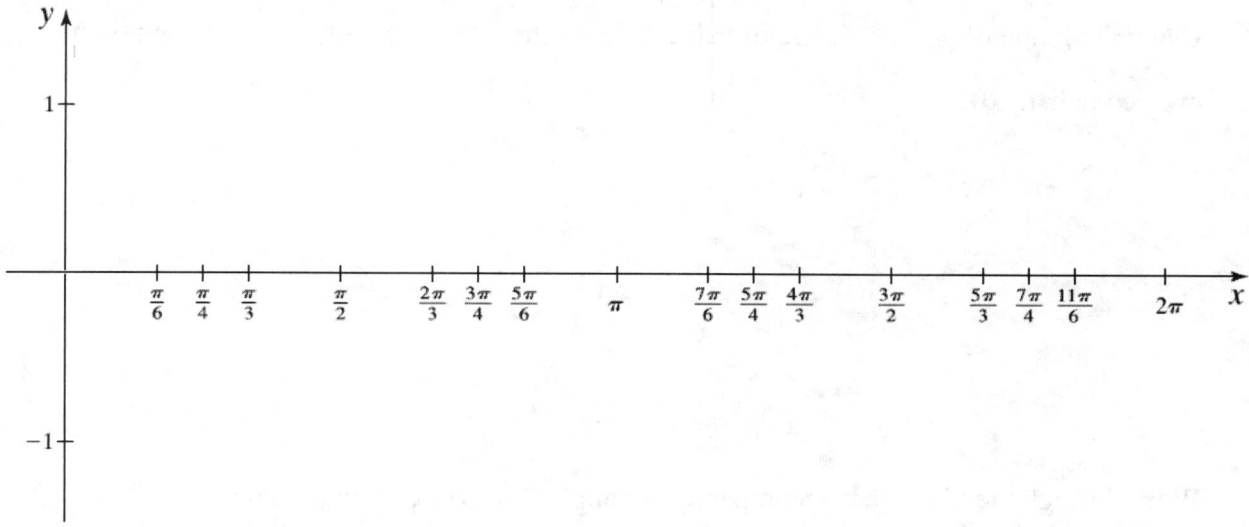

What is the definition of a **Periodic Function**?

Write down the **Characteristics of the Sine Function**.

Section 2.1

Work through the video that accompanies Example 1 showing all work below.

Using the graph of $y = \sin x$, list all values of $x$ on the interval $\left[-3\pi, \dfrac{7\pi}{4}\right]$ that satisfy the ordered pair $(x, 0)$.

Work through the video that accompanies Example 2 showing all work below.
Use the periodic property of $y = \sin x$ to determine which of the following expressions is equivalent to $\sin\left(\dfrac{23\pi}{6}\right)$.

i. $\sin\left(\dfrac{\pi}{6}\right)$  ii. $\sin\left(\dfrac{5\pi}{6}\right)$  iii. $\sin\left(\dfrac{11\pi}{6}\right)$  iv. $\sin\left(\dfrac{13\pi}{6}\right)$

Work through Example 3 showing all work below.
Use the fact that $y = \sin x$ is an odd function to determine which of the following expressions is equivalent to $-\sin\left(\dfrac{9\pi}{16}\right)$.

i. $\sin\left(-\dfrac{9\pi}{16}\right)$  ii. $\sin\left(\dfrac{9\pi}{16}\right)$  iii. $-\sin\left(-\dfrac{9\pi}{16}\right)$

Section 2.1

## Section 2.1 Objective 2 Understanding the Graph of the Cosine Function and Its Properties

Fill in the four tables below:

| $x$ | $0$ | $\frac{\pi}{2}$ | $\pi$ | $\frac{3\pi}{2}$ | $2\pi$ |
|---|---|---|---|---|---|
| $y = \cos x$ | | | | | |

| $x$ | $\frac{\pi}{6}$ | $\frac{5\pi}{6}$ | $\frac{7\pi}{6}$ | $\frac{11\pi}{6}$ |
|---|---|---|---|---|
| $y = \cos x$ | | | | |

| $x$ | $\frac{\pi}{4}$ | $\frac{3\pi}{4}$ | $\frac{5\pi}{4}$ | $\frac{7\pi}{4}$ |
|---|---|---|---|---|
| $y = \cos x$ | | | | |

| $x$ | $\frac{\pi}{3}$ | $\frac{2\pi}{3}$ | $\frac{4\pi}{3}$ | $\frac{5\pi}{3}$ |
|---|---|---|---|---|
| $y = \cos x$ | | | | |

Sketch the graph of $y = \cos x$ on the interval $[0, 2\pi]$.

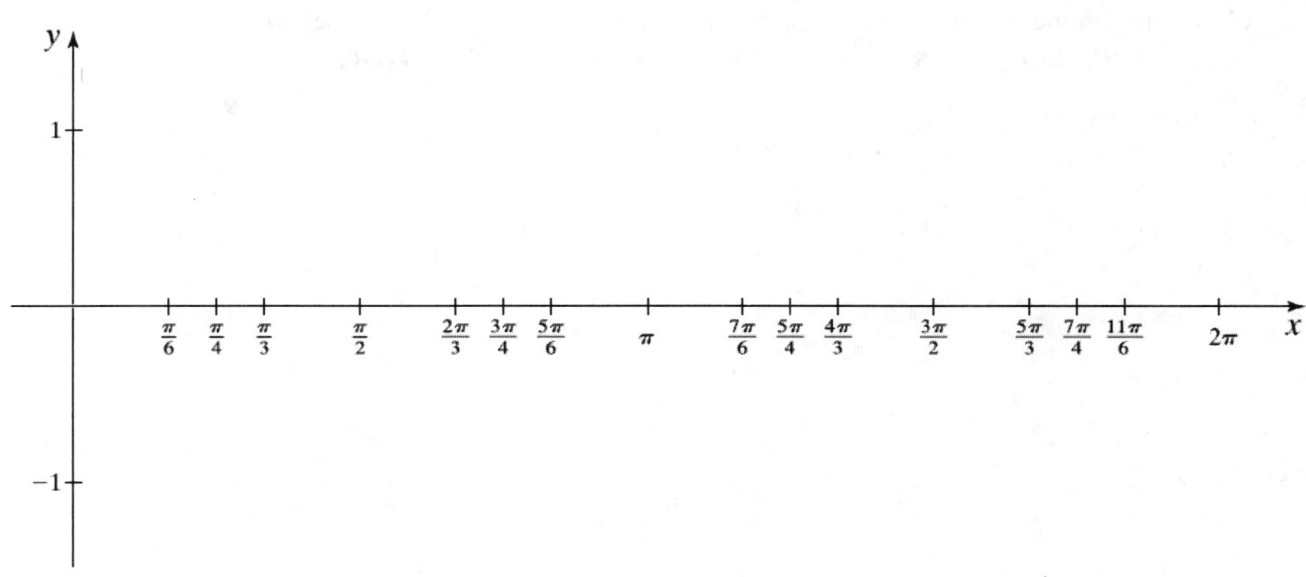

Section 2.1

Write down the **Characteristics of the Cosine Function.**

Work through the video that accompanies Example 4 showing all work below. Using the graph of $y = \cos x$, list all values of $x$ on the interval $[-\pi, 2\pi]$ that satisfy the ordered pair $\left(x, \dfrac{1}{2}\right)$.

Section 2.1

Write down the **Five Quarter Points** of $y = \sin x$ **and** $y = \cos x$ and plot them on the graphs below.  **You must memorize these quarter poimts!**

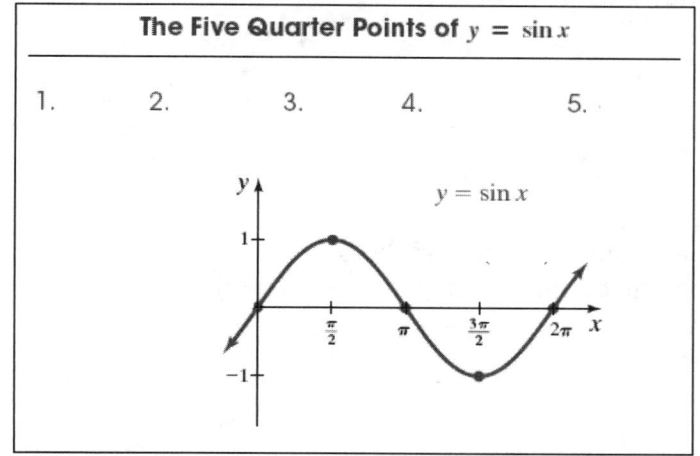

The Five Quarter Points of $y = \sin x$

1.     2.     3.     4.     5.

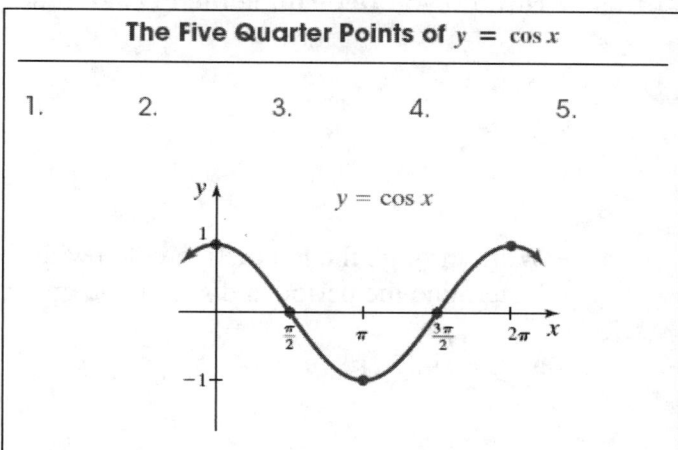

The Five Quarter Points of $y = \cos x$

1.     2.     3.     4.     5.

Section 2.1  Objective 3 Sketching Graphs of the From $y = A\sin x$ and $y = A\cos x$

What is the definition of **amplitude**?

Work through the video with Example 5 showing all work below.
Determine the amplitude and range of $y = -\dfrac{2}{3}\cos x$ and then sketch the graph.

69

Copyright © 2019 Pearson Education, Inc.

Section 2.1

Section 2.1 Objective 4 Sketching Graphs of the Form $y = \sin(Bx)$ and $y = \cos(Bx)$

How do you **Determine the Period of** $y = \sin(Bx)$ **and** $y = \cos(Bx)$?

Work through the interactive video with Example 6 and show all work below. Determine the period and sketch the graph of each function.

a. $y = \sin(2x)$

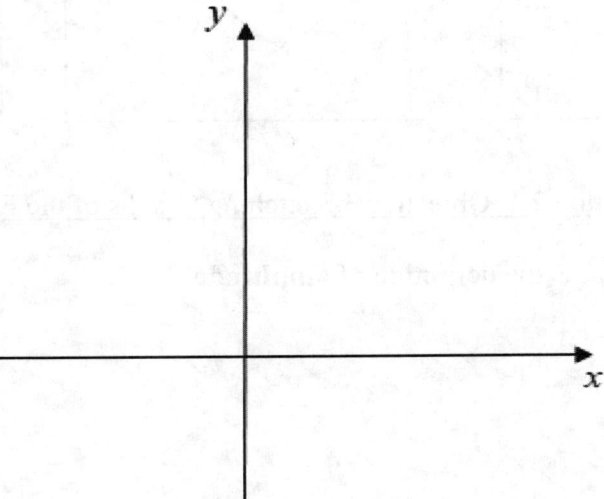

b. $y = \cos\left(\dfrac{1}{2}x\right)$

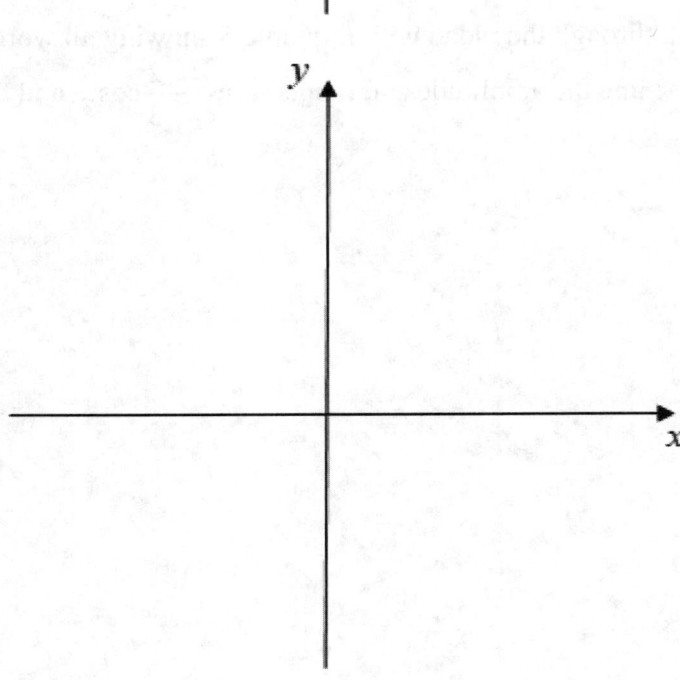

c. $y = \sin(\pi x)$

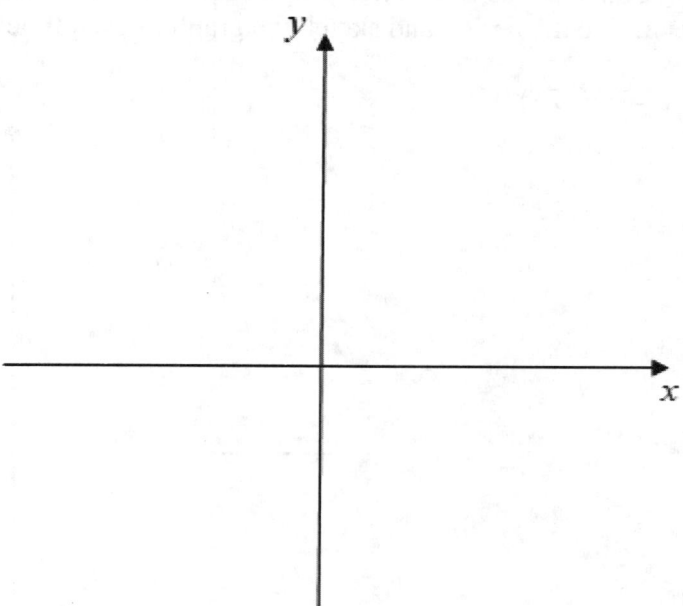

Turn to page 2.1-33 and fill in the blanks below:

Recall that $y = \sin x$ is an _____ function. This means that $\sin(-x) =$ _____ for all $x$ in the domain of $y = \sin x$. It follows that $\sin(-Bx) =$ _____ .

Recall that $y = \cos x$ is an _____ function. This means that $\cos(-x) =$ _____ for all $x$ in the domain of $y = \cos x$. It follows that $\cos(-Bx) =$ _____ .

Therefore, when $B > 0$, we can sketch the graph of $y = \sin(-Bx)$ by simply sketching the graph of _____ .

The graph of $y = \cos(-Bx)$ is the _____ graph as the graph of _____ .

Section 2.1

Work through the interactive video with Example 7 and show all work below.
Determine the period and sketch the graph of each function.

a. $y = \sin(-2x)$

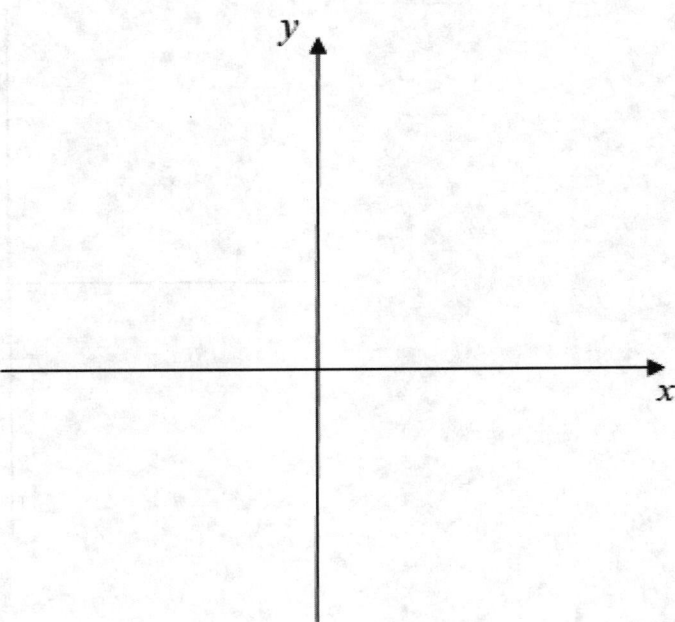

b. $y = \cos\left(-\dfrac{1}{2}x\right)$

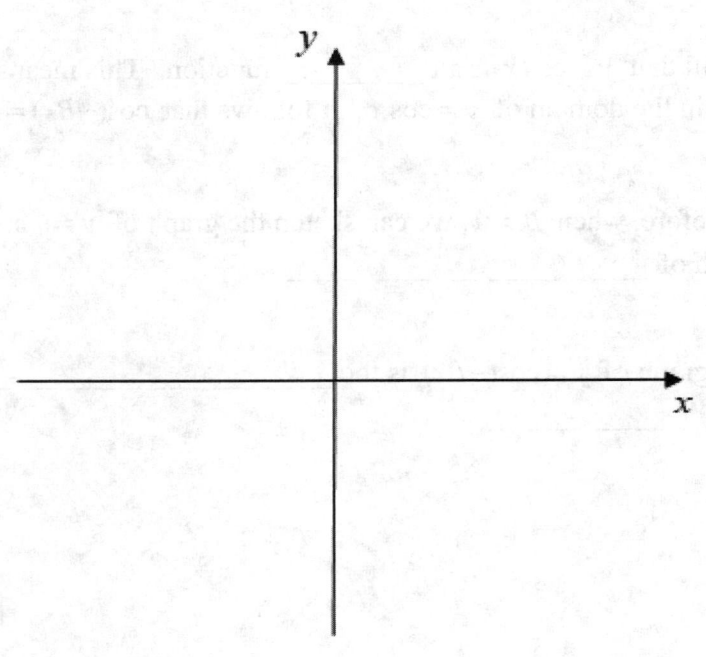

c. $y = \sin(-\pi x)$

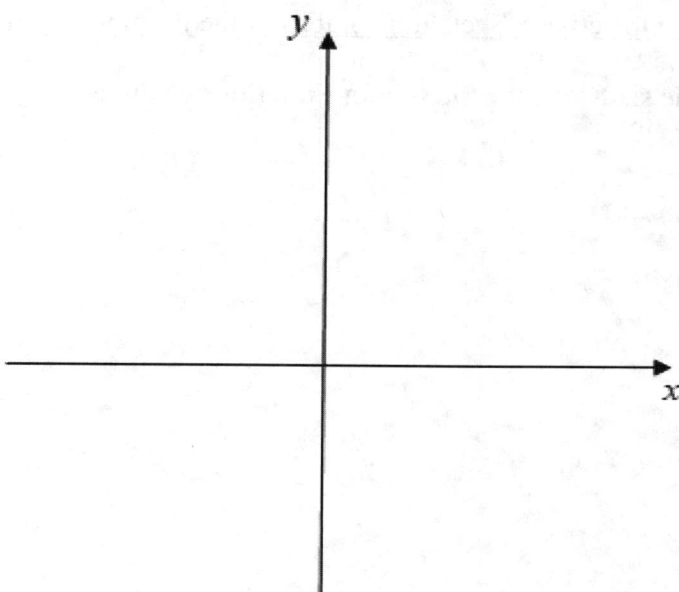

Section 2.1

## Section 2.1 Objective 5 Sketching Graphs of the Form $y = A\sin(Bx)$ and $y = A\cos(Bx)$

What are the six **Steps for Sketching Functions of the Form $y = A\sin(Bx)$ and $y = A\cos(Bx)$**?

**Step 1.**

**Step 2.**

**Step 3.**

**Step 4.**

**Step 5.**

**Step 6.**

Work through the interactive video with Example 8 and show all work below.
Use the six-step process outlined in this section to sketch each graph.

a. $y = 3\sin(4x)$

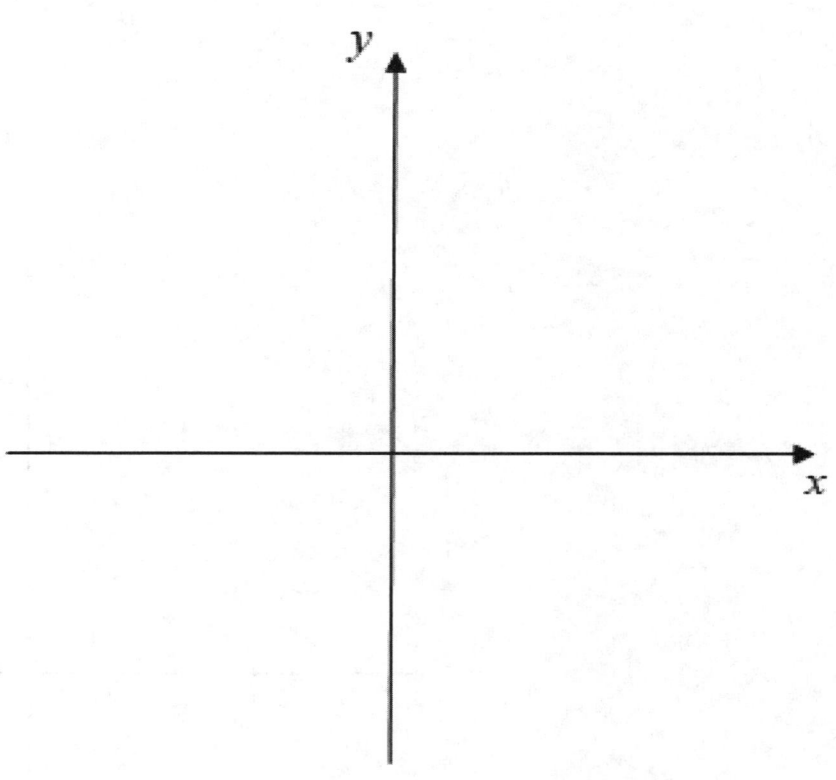

b. $y = -2\cos\left(\dfrac{1}{3}x\right)$

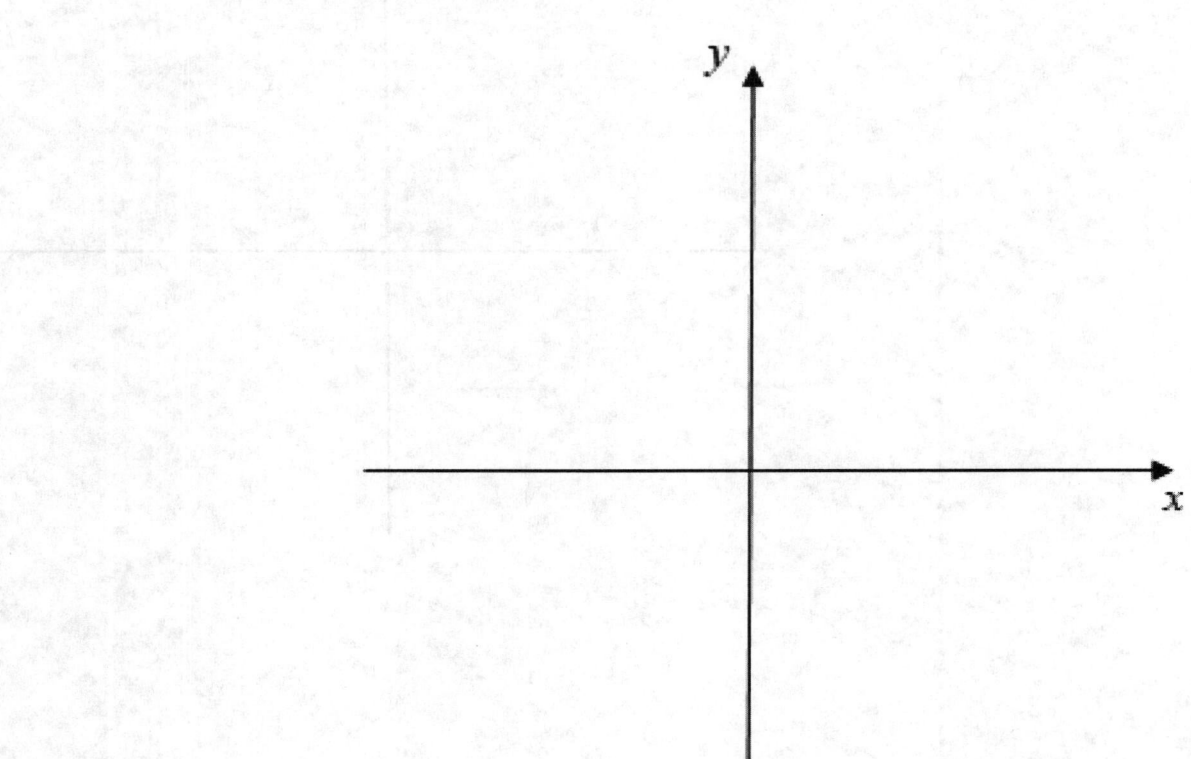

c. $y = -6\sin\left(-\dfrac{\pi x}{2}\right)$

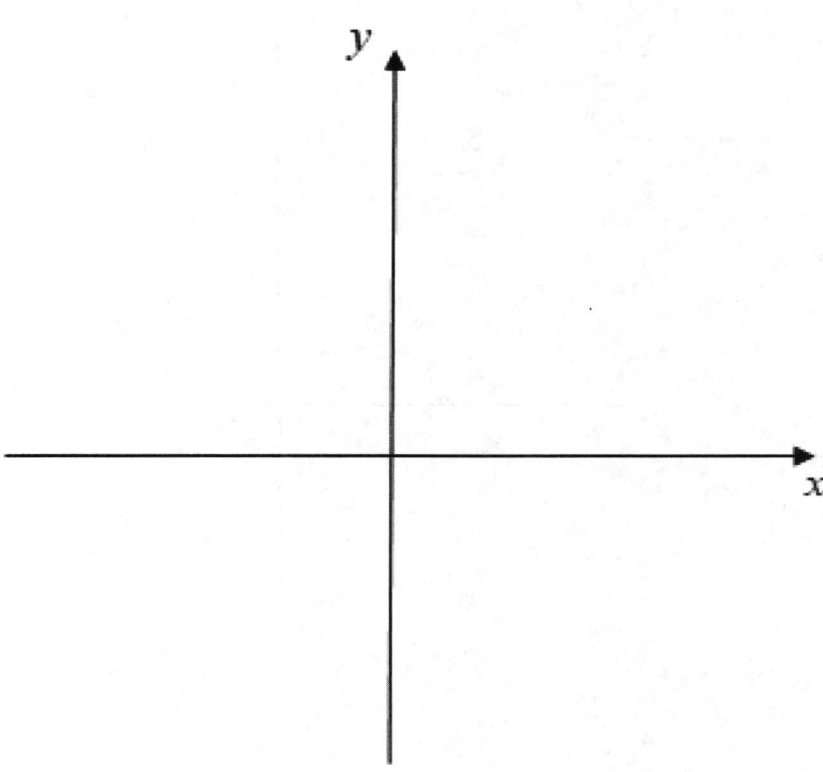

Section 2.1

Work through the **Guided Visualization** seen on page 2.1-46. Sketch and label the graphs of $y = A\sin(Bx)$ and $y = A\cos(Bx)$. Make sure that at least one of your graphs has $A < 0$.

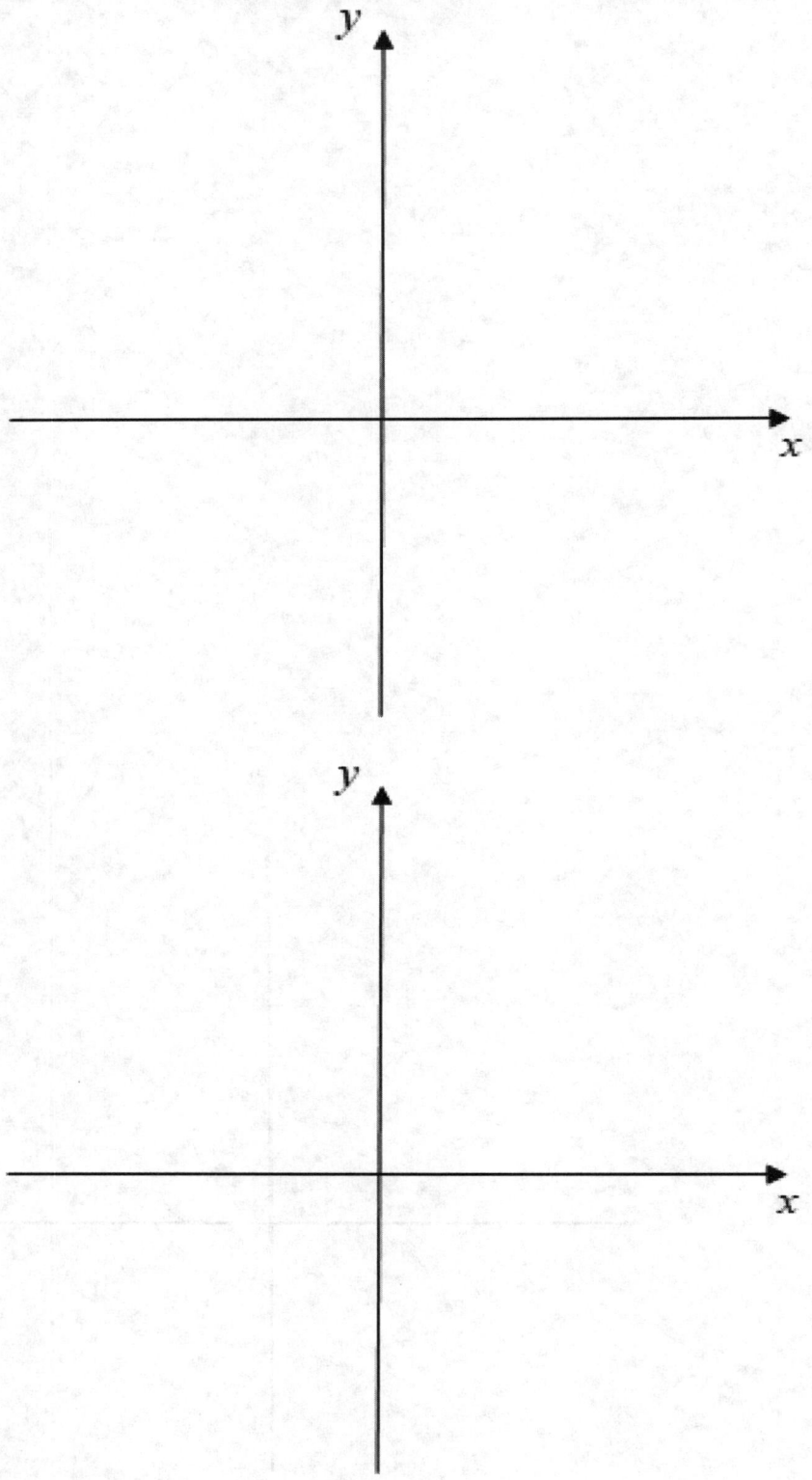

Section 2.1

Section 2.1  Objective 6 Determine the Equation of a Function of the Form $y = A\sin(Bx)$ and $y = A\cos(Bx)$ Given Its Graph

Given the graph of a function of the form $y = A\sin(Bx)$ or $y = A\cos(Bx)$, what three things must we determine in order to determine the proper function?

1.

2.

3.

Write the three steps for Determining the Equation of a Function of the form $y = A\sin(Bx)$ or $y = A\cos(Bx)$ Given the Graph.

**Step 1.**

**Step 2.**

**Step 3.**

Section 2.1

Work through the interactive video that accompanies Example 9.
One cycle of the graphs of three trigonometric functions of the form $y = A\sin(Bx)$ or $y = A\cos(Bx)$ for $B > 0$ are given below. Determine the equation of the function represented by each graph.

a.

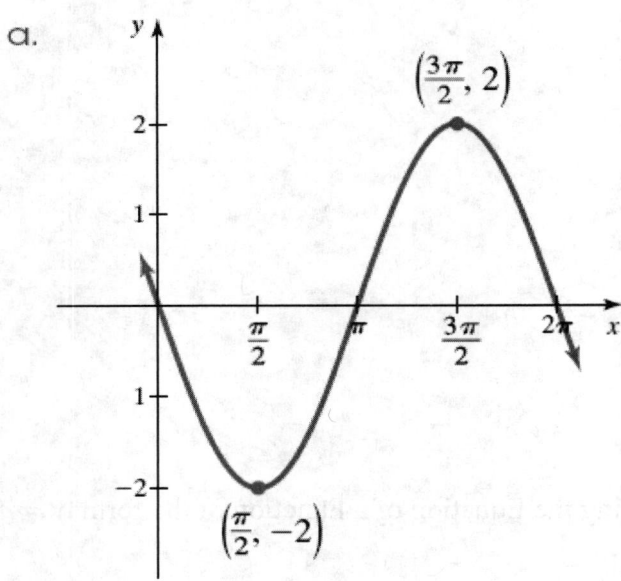

b.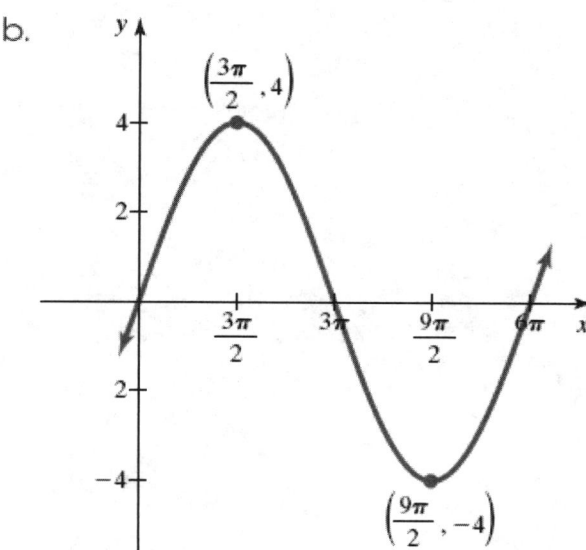

c.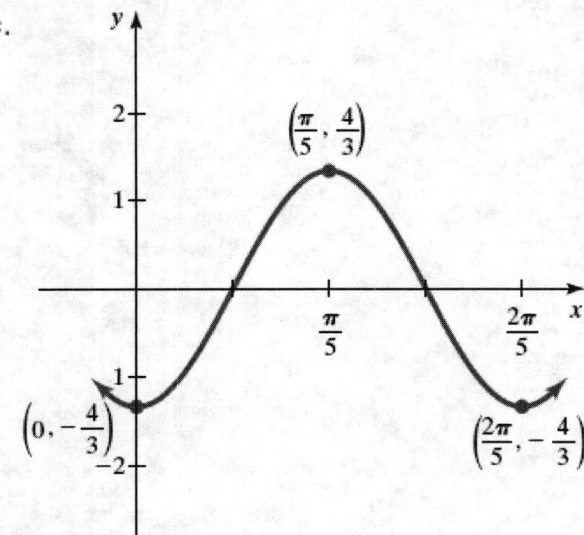

## Section 2.2 Guided Notebook

**2.2 More on Graphs of Sine and Cosine: Phase Shift**
- ☐ Work through Section 1.2 TTK #1–5
- ☐ Work through Section 1.2 Objective 1
- ☐ Work through Section 1.2 Objective 2
- ☐ Work through Section 1.2 Objective 3

### Section 2.2 More on Graphs of Sine and Cosine: Phase Shift

### 2.2 Things To Know

1. Using Vertical Shifts to Graph Functions
Try working through a "You Try It" problem or watch the animation.




2. Using Horizontal Shifts to Graph Functions
Try working through a "You Try It" problem or watch the animation.




3. Sketching Graphs of the Form $y = A\sin x$ and $y = A\cos x$
Try working through a "You Try It" problem or watch the video.

Section 2.2

4. Sketching Graphs of the Form $y = \sin(Bx)$ and $y = \cos(Bx)$
Try working through a "You Try It" problem or watch the interactive video.

5. Sketching Graphs of the Form $y = A\sin(Bx)$ and $y = A\cos(Bx)$
Try working through a "You Try It" problem or watch the interactive video.

Section 2.2  Objective 1 Sketching Graphs of the Form $y = \sin(x-C)$ and $y = \cos(x-C)$

Use the graphs of $y = \sin x$ and $y = \cos x$ on the left, to sketch the graphs of $y = \sin(x-C)$, $y = \sin(x+C)$, $y = \cos(x-C)$, and $y = \cos(x+C)$ for $C > 0$.

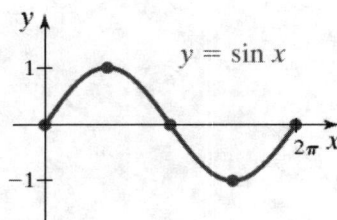
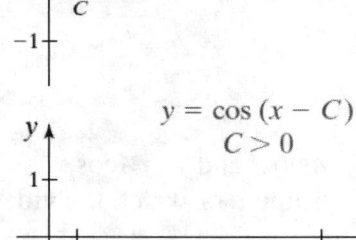
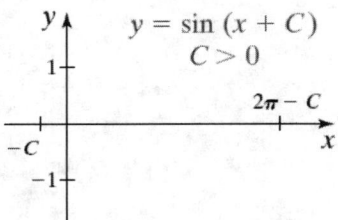
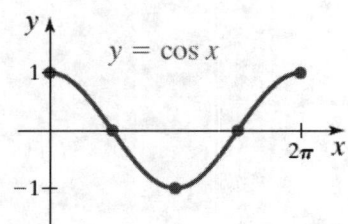
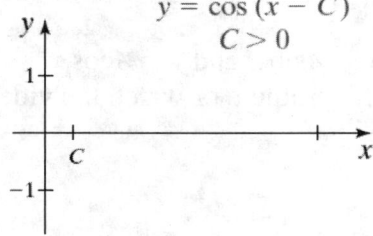
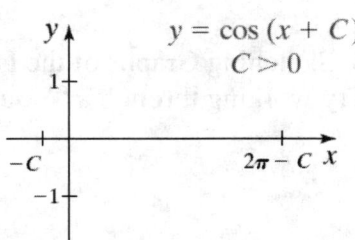

Section 2.2

- What is the **phase shift**?

- Write down the properties and sketch the graphs of $y = \sin(x - C)$ and $y = \cos(x - C)$.

Work through the interactive video with Example 1 showing all work below. Determine the phase shift and sketch the graph of each function.

a. $y = \cos(x - \pi)$

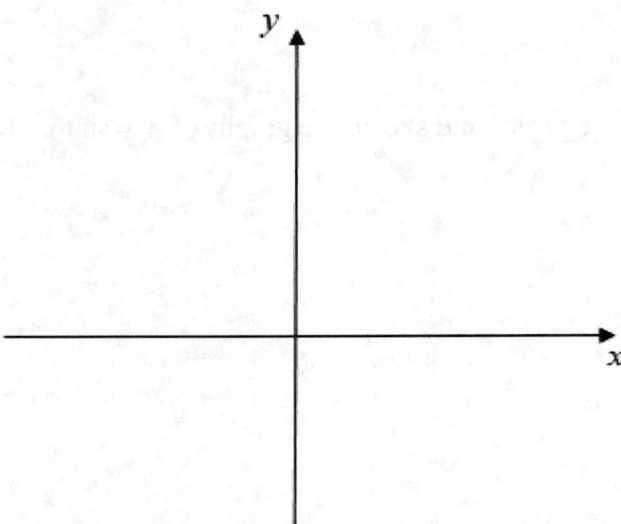

b. $y = \sin\left(x + \dfrac{\pi}{2}\right)$

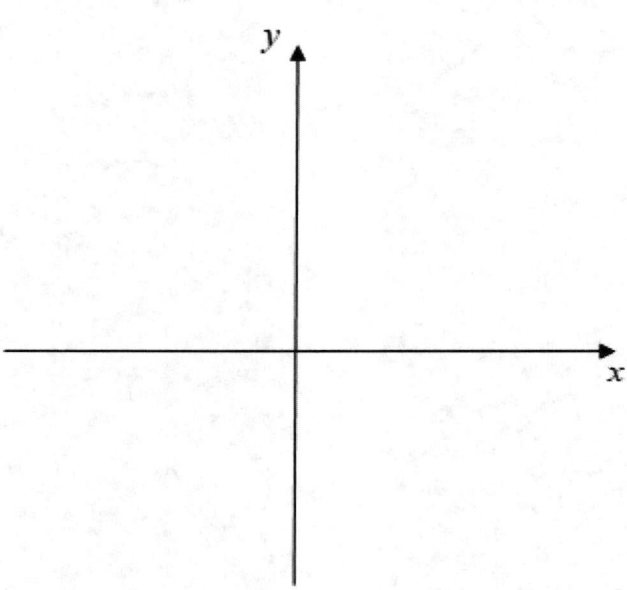

What is the **Relationship between Graphs of Sine and Cosine Functions?**

Fill in the blanks:

$\sin\left(x+\dfrac{\pi}{2}\right) =$ _____ and $\cos\left(x-\dfrac{\pi}{2}\right) =$ _____

Section 2.2 Objective 2 Sketching Graphs of the Form $y = A\sin(Bx - C)$ and $y = A\cos(Bx - C)$

What are the seven **Steps for Sketching Functions of the Form** $y = A\sin(Bx - C)$ **and** $y = A\cos(Bx - C)$?

**Step 1.**

**Step 2.**

**Step 3.**

**Step 4.**

**Step 5.**

**Step 6.**

**Step 7.**

Section 2.2

Work through the interactive video with Example 2 showing all work below. Sketch the graph of each function.

a. $y = 3\sin(2x - \pi)$

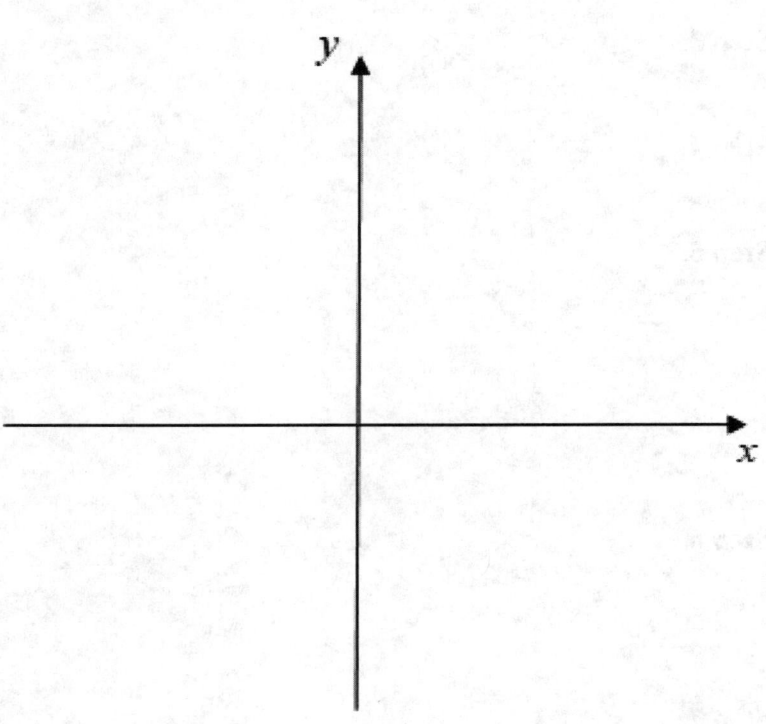

b. $y = -2\cos\left(3x + \dfrac{\pi}{2}\right)$

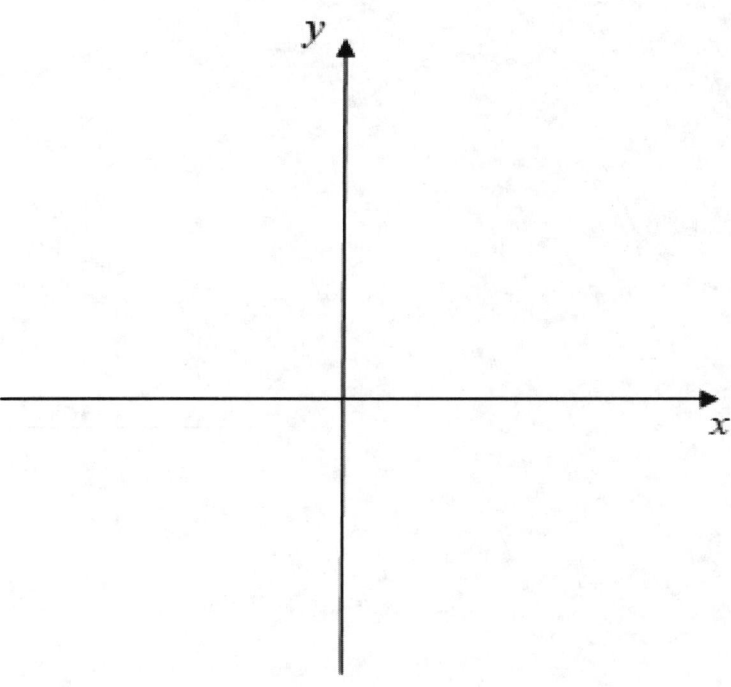

Section 2.2

c. $y = 3\sin(\pi - 2x)$

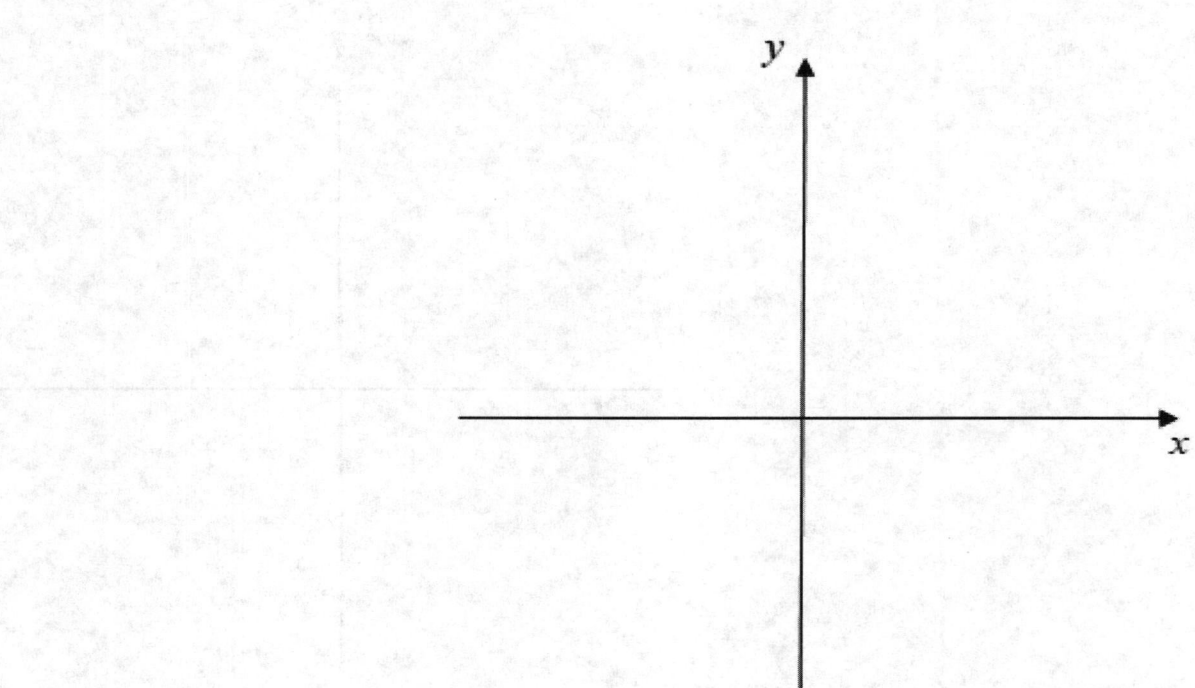

d. $y = -2\cos\left(-3x + \dfrac{\pi}{2}\right)$

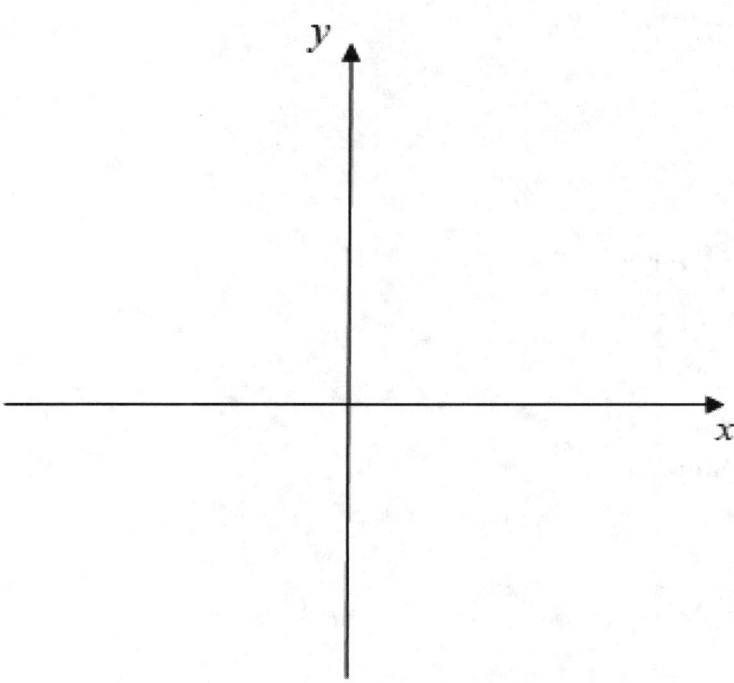

Section 2.2

Section 2.2  Objective 3 Sketching Graphs of the Form $y = A\sin(Bx - C) + D$ and $y = A\cos(Bx - C) + D$

What are the seven **Steps for Sketching Functions of the Form** $y = A\sin(Bx - C) + D$ **and** $y = A\cos(Bx - C) + D$?

**Step 1.**

**Step 2.**

**Step 3.**

**Step 4.**

**Step 5.**

**Step 6.**

**Step 7.**

Work through the interactive video with Example 3 showing all work below. Sketch the graph of each function.

a. $y = 3\sin\left(2x - \dfrac{\pi}{2}\right) - 1$

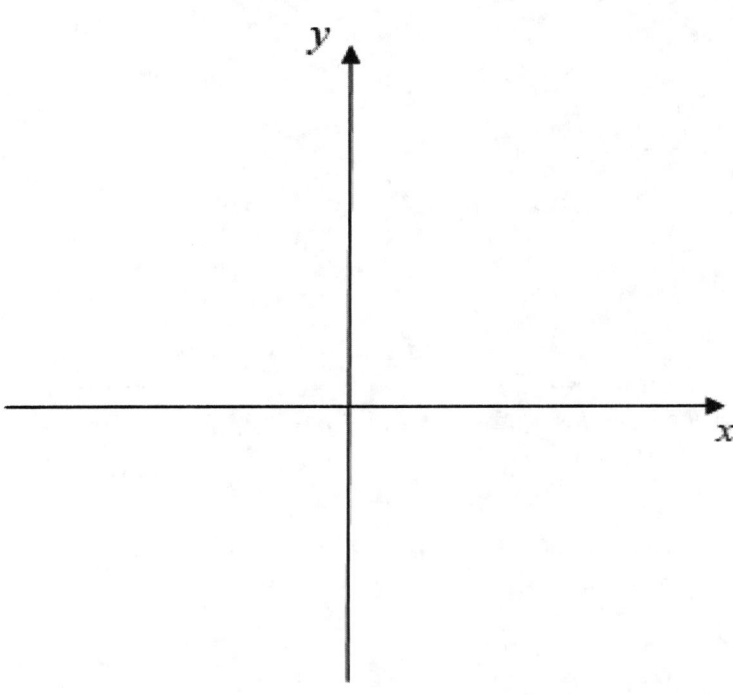

Section 2.2

b. $y = 4 - \cos(-\pi x + 2)$

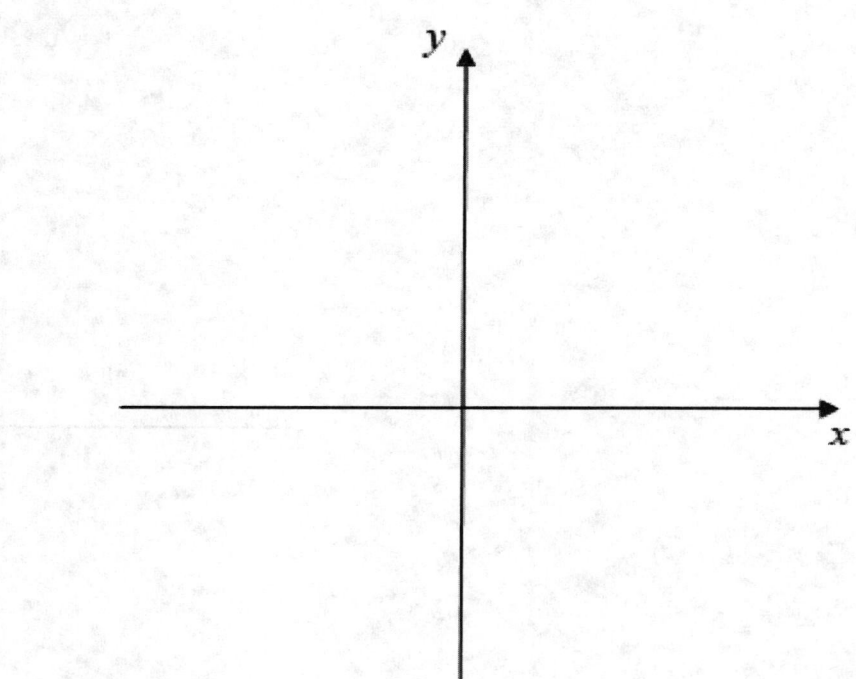

Work through the **Guided Visualization** seen on page 1.2-43. Sketch and label the graphs of $y = A\sin(Bx - C) + D$ and $y = A\cos(Bx - C) + D$. You choose the values of $A$, $B$, $C$, and $D$.

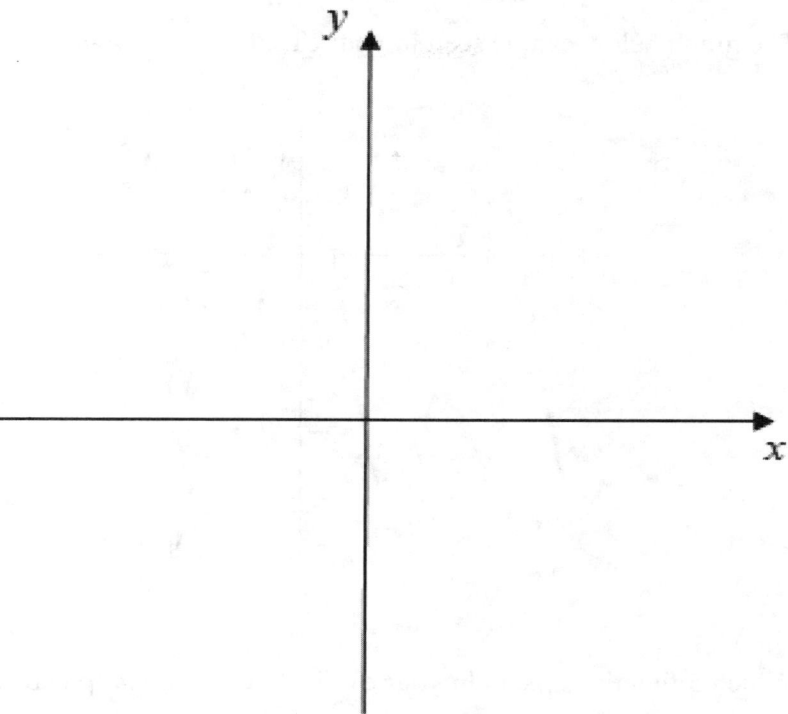

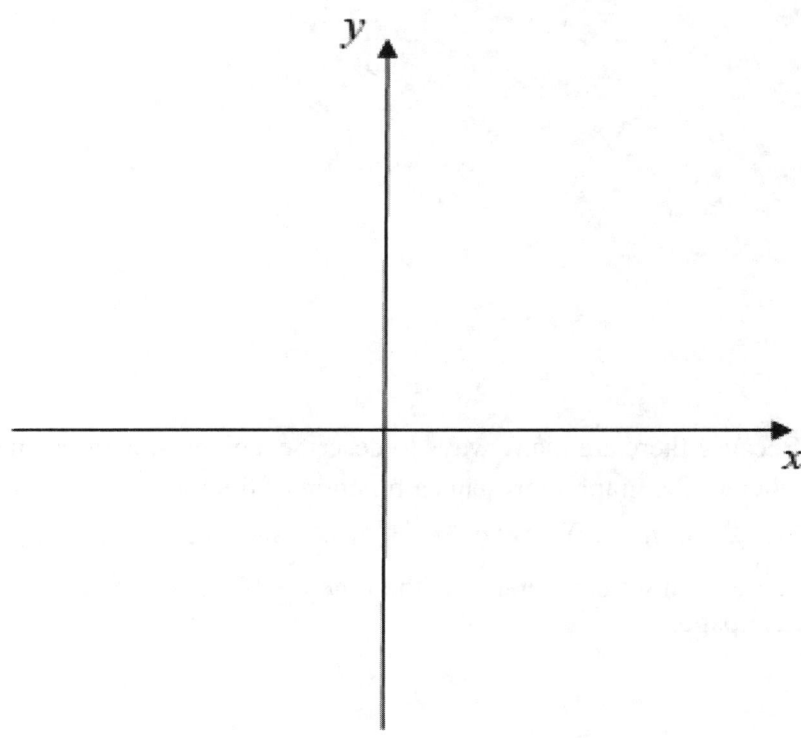

Section 2.2

## Section 2.2 Objective 4 Determine the Equation of a Function of the Form $y = A\sin(Bx - C) + D$ and $y = A\cos(Bx - C) + D$ Given Its Graph

The graph below can be seen in your eText.

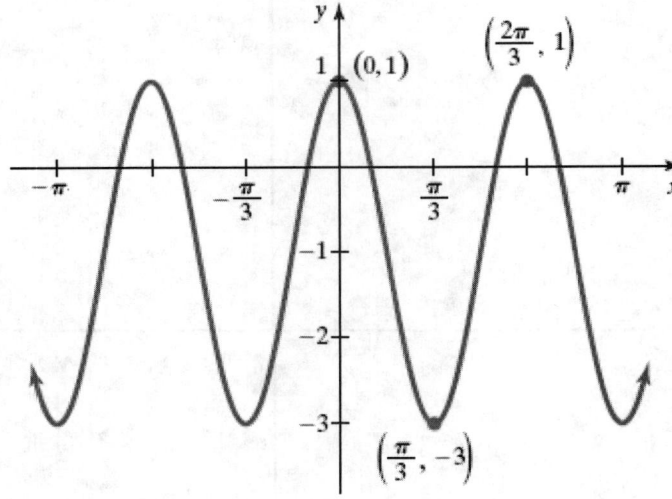

Which 5 functions listed in your eText describe the graph above?

Because there are many ways to describe a given sine or cosine curve, we must be given whether the graph represents a function of the form $y = A\sin(Bx - C) + D$ or of the form $y = A\cos(Bx - C) + D$. We will always also assume that $B > 0$. If these assumptions are made, then we can determine the function by following the six steps that can be seen on the next page.

Section 2.2

Write down the six steps necessary for determining an equation of a function of the form $y = A\sin(Bx - C) + D$ or $y = A\cos(Bx - C) + D$ given the graph.

**Step 1.**

**Step 2.**

**Step 3.**

**Step 4.**

**Step 5.**

**Step 6.**

Section 2.2

Work through the interactive video that accompanies Example 4.

Example 4a.

The graph of a function of the form $y = A\cos(Bx - C) + D,$ where $B > 0$ is given below.

The five quarter points of one cycle of the graph are labeled. These five quarter points on the graph correspond to the five quarter points of the graph of $y = \cos x$ over the interval $[0, 2\pi]$.

Determine the specific function that is represented by the given graph based on the association of the labeled quarter points and the quarter points of the graph of $y = \cos x$ over the interval $[0, 2\pi]$.

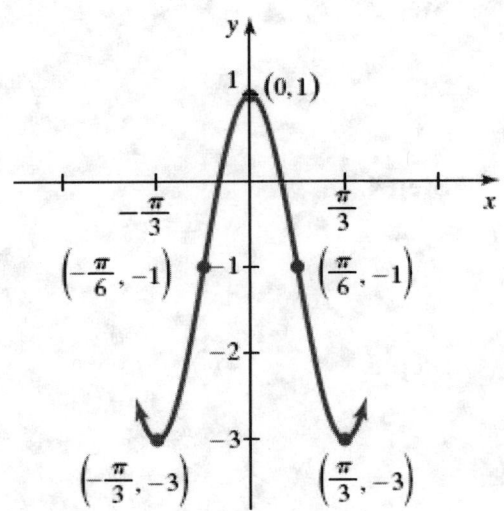

Example 4b.

The graph of a function of the form $y = A\sin(Bx - C) + D$ where $B > 0$ is given below. The five quarter points of one cycle of the graph are labeled. These five quarter points on the graph correspond to the five quarter points of the graph of $y = \sin x$ over the interval $[0, 2\pi]$. Determine the specific function that is represented by the given graph based on the association of the labeled quarter points and the quarter points of the graph of $y = \sin x$ over the interval $[0, 2\pi]$.

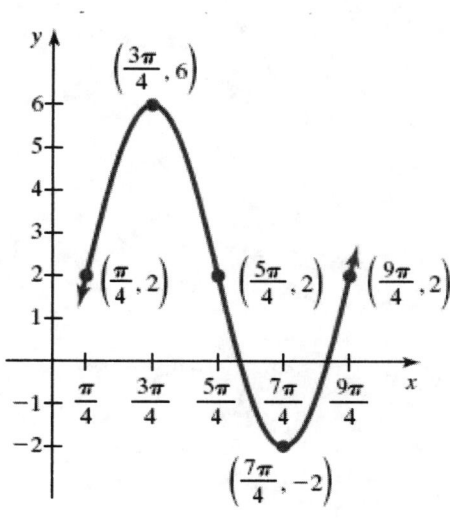

99

## Section 2.3 Guided Notebook

**2.3 The Graphs of the Tangent, Cotangent, Secant, and Cosecant Functions**
- [ ] Work through Section 2.3 TTK #1–5
- [ ] Work through Section 2.3 Objective 1
- [ ] Work through Section 2.3 Objective 2
- [ ] Work through Section 2.3 Objective 3
- [ ] Work through Section 2.3 Objective 4
- [ ] Work through Section 2.3 Objective 5
- [ ] Work through Section 2.3 Objective 6
- [ ] Work through Section 2.3 Objective 7

## 2.3 Things To Know

1. Using the Special Right Triangle
Try working through a "You Try It" problem or watch the video.

2. Finding the Values of the Trigonometric Functions of Quadrantal Angles
Try working through a "You Try It" problem or watch the video.

3. Evaluating Trigonometric Functions of Angles Belonging to the $\frac{\pi}{3}$, $\frac{\pi}{6}$, or $\frac{\pi}{4}$ Families
Try working through a "You Try It" problem or watch the interactive video.

Section 2.3

4. Sketching Graphs of the Form $y = A\sin(Bx)$ and $y = A\cos(Bx)$
Try working through a "You Try It" problem or watch the interactive video.

5. Sketching Graphs of the Form $y = A\sin(Bx - C) + D$ and $y = A\cos(Bx - C) + D$
Try working through a "You Try It" problem or watch the interactive video.

Section 2.3 Objective 1 Understanding the Graph of the Tangent Function and Its Properties

Watch the video that accompanies this objective to see how to sketch the graph of $y = \tan x$.

Sketch the graph of the **principal cycle** of $y = \tan x$.

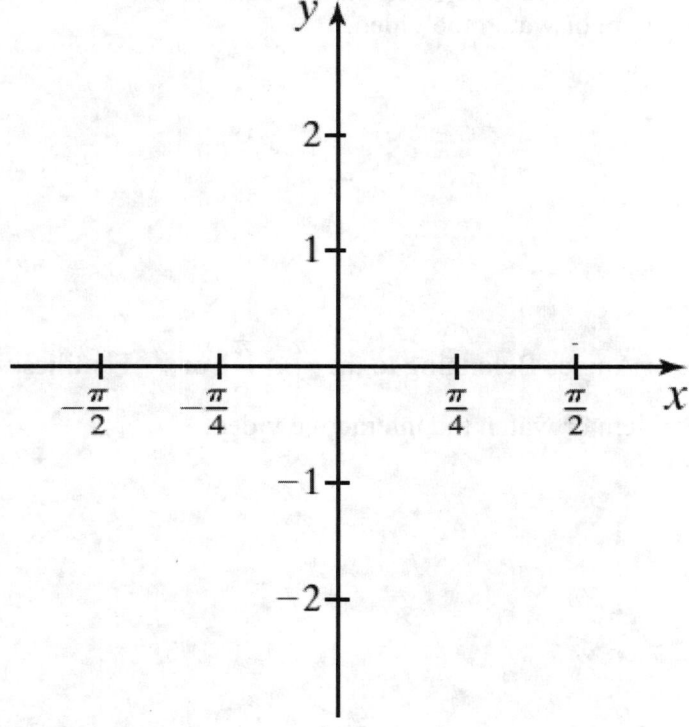

Section 2.3

What are the three special points in each cycle of the graph of $y = \tan x$ that will help us to sketch the graph? Go back to the sketch above and add these special points. Also, make sure to draw and label the vertical asymptotes.

Fill in the blanks below:

The principal cycle of the tangent function is defined on the interval: _____

The coordinates of the center point of the principal cycle of the tangent function are _____.

The coordinates of the two halfway points are: _____ and _____.

Write down the **Characteristics of the Tangent Function.**

Section 2.3

Work through the video with Example 1 showing all work below.
List all the halfway points of $y = \tan x$ on the interval $\left[-\pi, \dfrac{5\pi}{2}\right]$ that have a $y$-coordinate of $-1$.

Section 2.3

Section 2.3 Objective 2 Sketching Functions of the Form $y = A\tan(Bx - C) + D$

What are the six **Steps for Sketching Functions of the Form** $y = A\tan(Bx - C) + D$?

**Step 1.**

**Step 2.**

**Step 3.**

**Step 4.**

**Step 5.**

**Step 6.**

Section 2.3

Work through the interactive video with Example 2 showing all work below.
For each function, determine the interval for the principal cycle. Then for the principal cycle, determine the equations of the vertical asymptotes, the coordinates of the center points, and the coordinates of the halfway points. Sketch the graph.

a.  $y = \tan\left(x - \dfrac{\pi}{6}\right)$

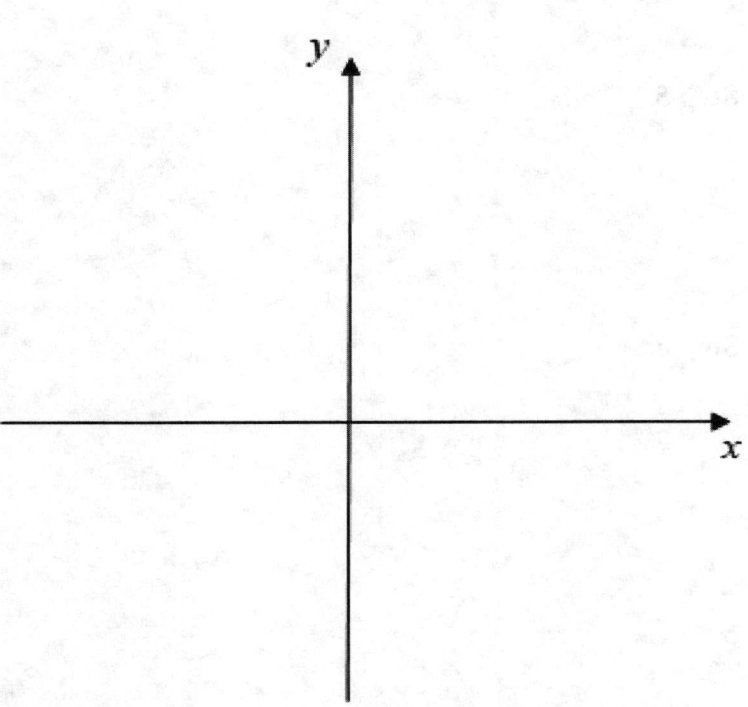

b. $y = 4\tan(\pi - 2x) + 3$

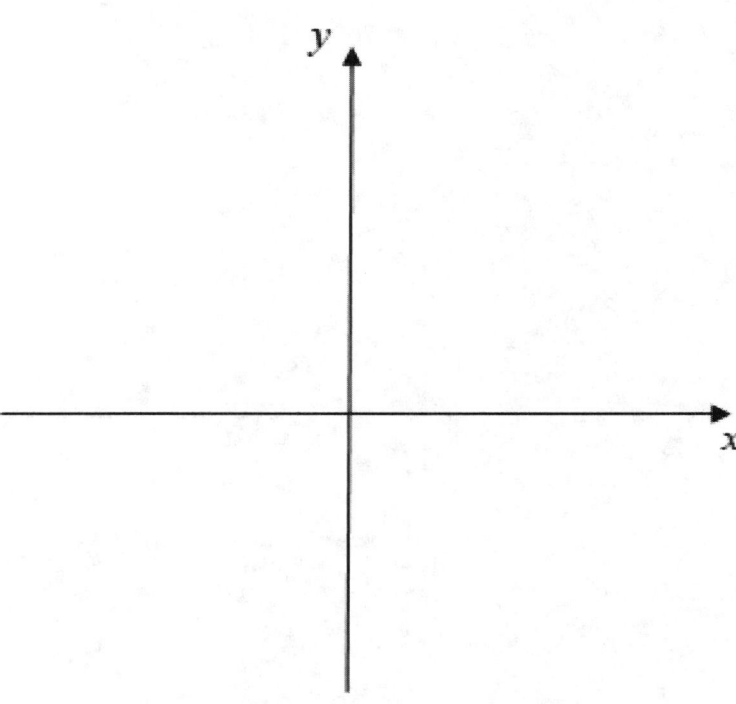

c. $y = \dfrac{1}{2}\tan(3x) - 1$

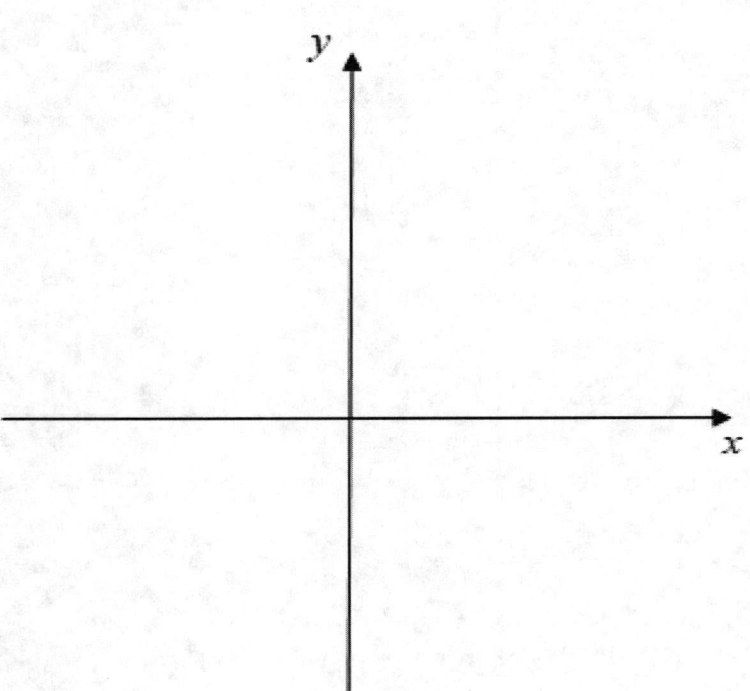

Section 2.3

## Section 2.3 Objective 3 Understanding the Graph of the Cotangent Function and Its Properties

Watch the video that accompanies this objective to see how to sketch the graph of $y = \cot x$.

Sketch the graph of the **principal cycle** of $y = \cot x$.

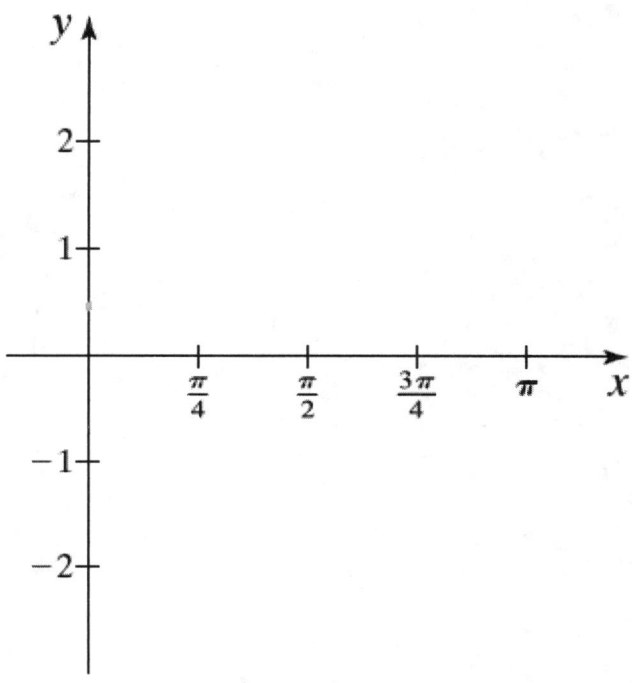

What are the three special points in each cycle of the graph of $y = \cot x$ that will help us to sketch the graph? Go back to the sketch above and add these special points. Also, make sure to draw and label the vertical asymptotes.

Fill in the blanks below:

The principal cycle of the cotangent function is defined on the interval: _____.

The coordinates of the center point of the principal cycle of the cotangent function are _____.

The coordinates of the two halfway points are: _____ and _____.

Section 2.3

What are the **Characteristics of the Cotangent Function**?

Work through the video with Example 3 showing all work below.
List all points on the graph of $y = \cot x$ on the interval $[-2\pi, 2\pi]$ that have a $y$-coordinate of $-\sqrt{3}$.

Section 2.3

Section 2.3 Objective 4 Sketching Functions of the Form $y = A\cot(Bx - C) + D$

What are the six **Steps for Sketching Functions of the Form** $y = A\cot(Bx - C) + D$?

**Step 1.**

**Step 2.**

**Step 3.**

**Step 4.**

**Step 5.**

**Step 6.**

Section 2.3

Work through the interactive video with Example 4 and show all work below.
For each function, determine the interval for the principal cycle. Then for the principal cycle, determine the equations of the vertical asymptotes, the coordinates of the center points, and the coordinates of the halfway points. Sketch the graph.

a. $y = \cot(2x + \pi) + 1$

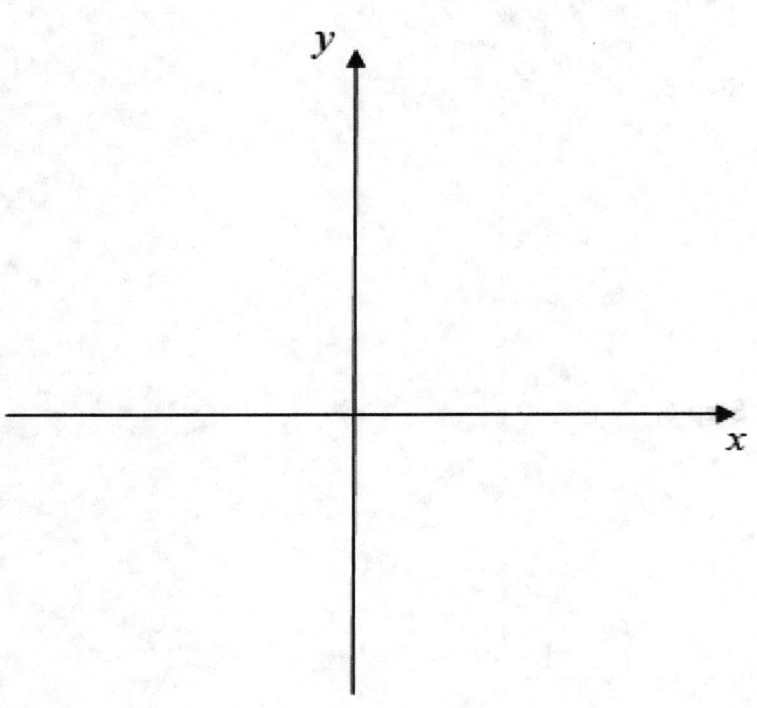

b. $y = -3\cot\left(x - \dfrac{\pi}{4}\right)$

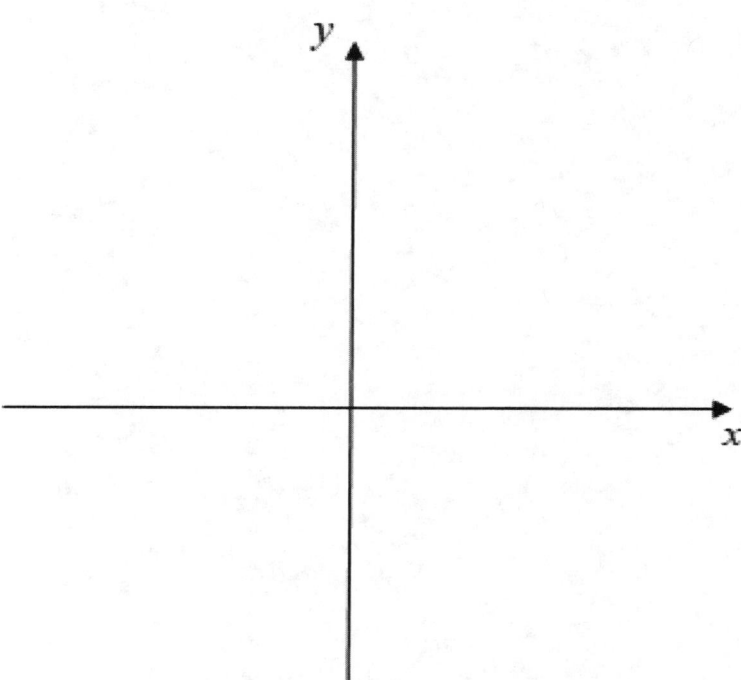

Section 2.3

## Section 2.3 Objective 5 Determine the Equation of a Function of the Form $y = A\tan(Bx - C) + D$ or $y = A\cot(Bx - C) + D$ Given Its Graph

Given a graph of the form $y = A\tan(Bx-C)+D$ or $y = A\cot(Bx-C)+D$ for $B > 0$, what five characteristics of the principal cycle of the given graph must we first identify in order to determine the proper function?

1.

2.

3.

4.

5.

Section 2.3

Write down the five steps for determining the equation of a function of the form $y = A\tan(Bx - C) + D$ or $y = A\cot(Bx - C) + D$ given the graph.

**Step 1.**

**Step 2.**

**Step 3.**

**Step 4.**

**Step 5.**

Section 2.3

Work through the interactive video that accompanies Example 5.
The principal cycle of the graphs of two trigonometric functions of the form
$y = A\tan(Bx - C) + D$ or $y = A\cot(Bx - C) + D$ for $B > 0$ are given below.
Determine the equation of the function represented by each graph.

a. $\quad y = A\tan(Bx - C) + D$

b. $y = A\cot(Bx - C) + D$

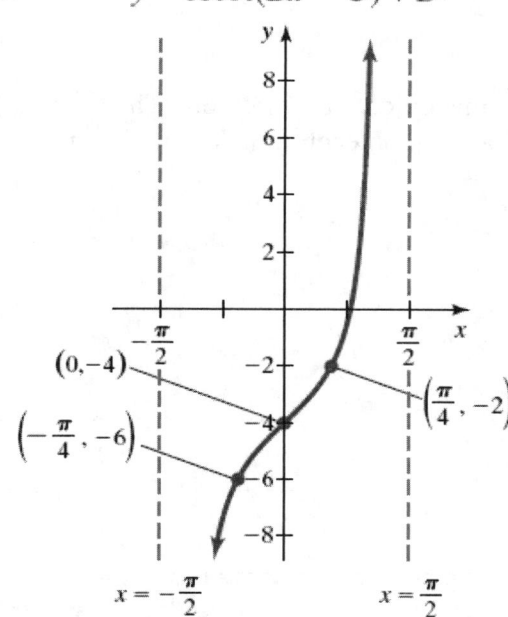

## Section 2.3 Objective 6 Understanding the Graphs of the Cosecant and Secant Functions and Their Properties

Work through the interactive video that accompanies this objective. Click on "The graph of $y = \csc x$" then click on "The graph of $y = \sec x$" and graph each function on the grids below.

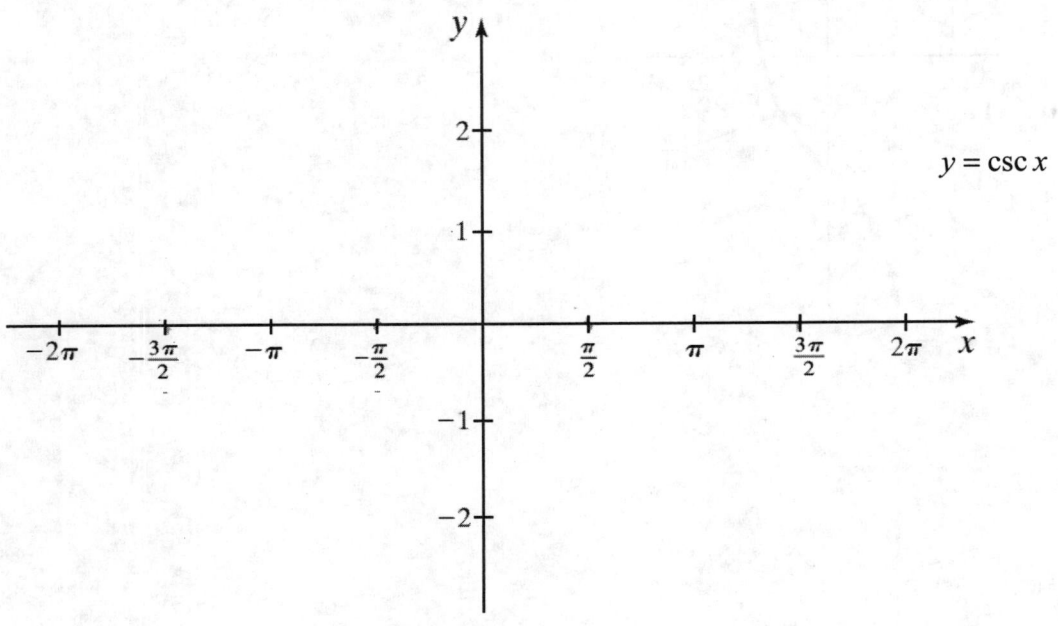

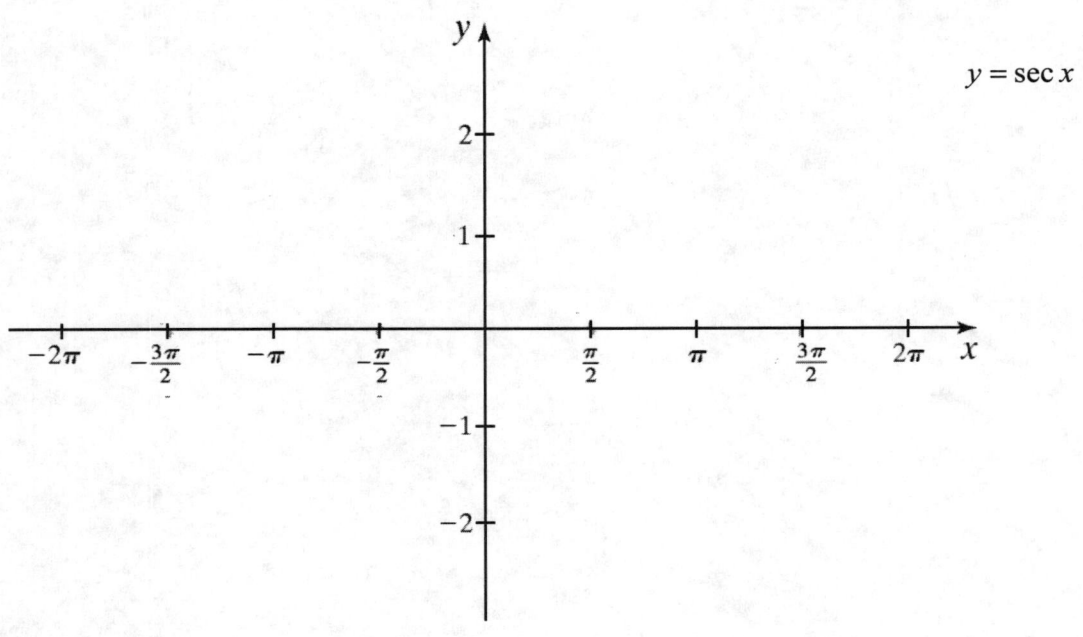

Section 2.3

- What are the **Characteristics of the Cosecant Function**?

- What are the **Characteristics of the Secant Function**?

Section 2.3

Section 2.3 Objective 7 Sketching Functions of the Form $y = A\csc(Bx - C) + D$ and $y = A\sec(Bx - C) + D$

What are the four **Steps for Sketching Functions of the Form** $y = A\csc(Bx - C) + D$ and $y = A\sec(Bx - C) + D$?

**Step 1.**

**Step 2.**

**Step 3.**

**Step 4.**

Section 2.3

Work through the interactive video with Example 6 and show all work below. Determine the equations of the vertical asymptotes and all relative maximum and relative minimum points of two cycles of each function and then sketch its graph.

a. $y = -2\csc(x+\pi)$

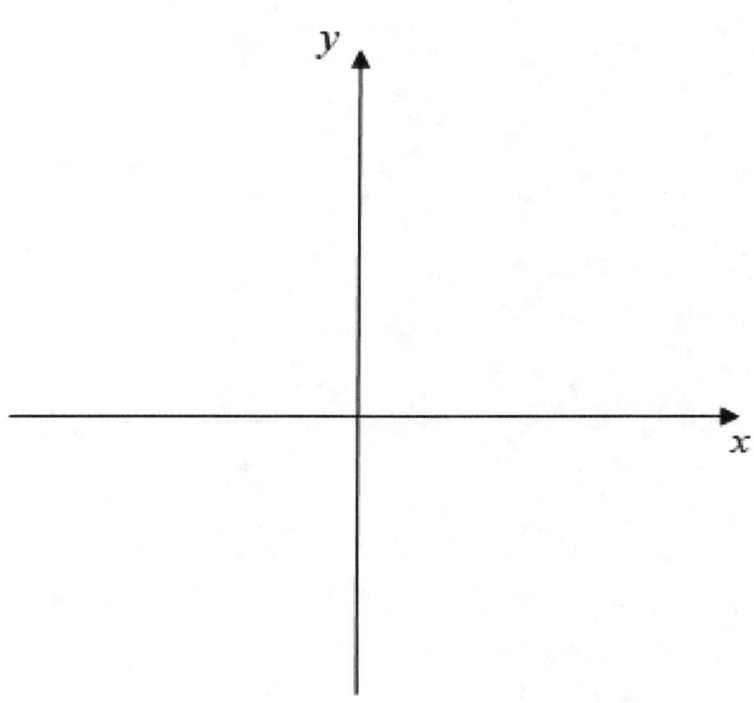

Section 2.3

b. $y = -\csc(\pi x) - 2$

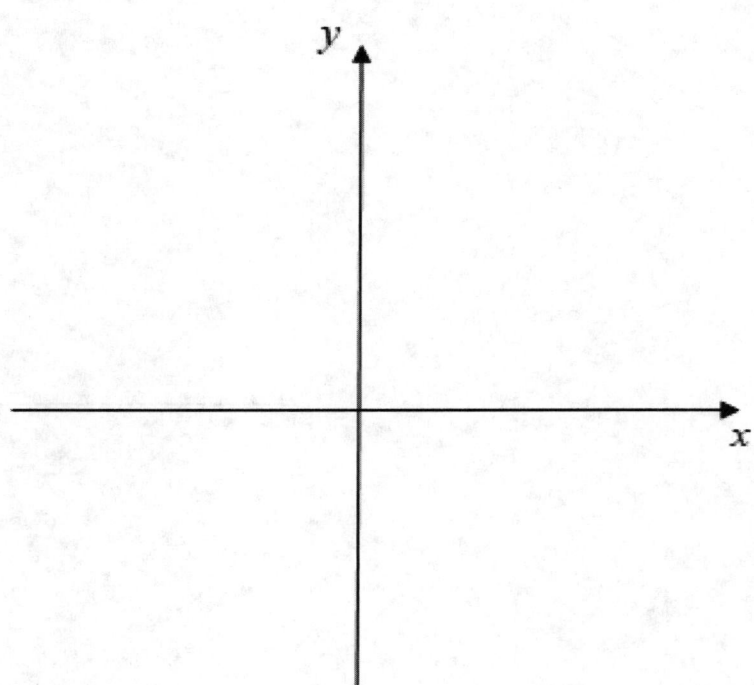

c. $y = 3\sec(\pi - x)$

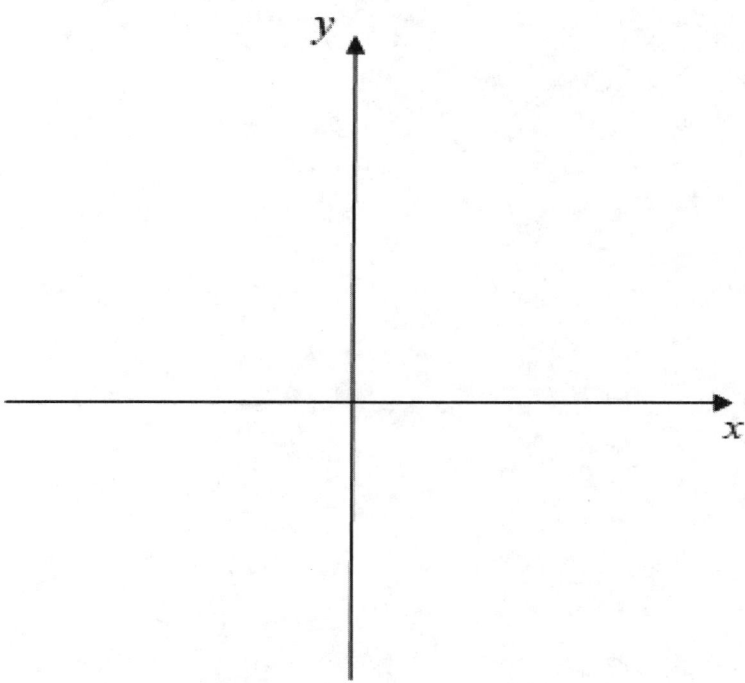

Section 2.3

d. $y = \sec\left(2x + \dfrac{\pi}{2}\right) + 1$

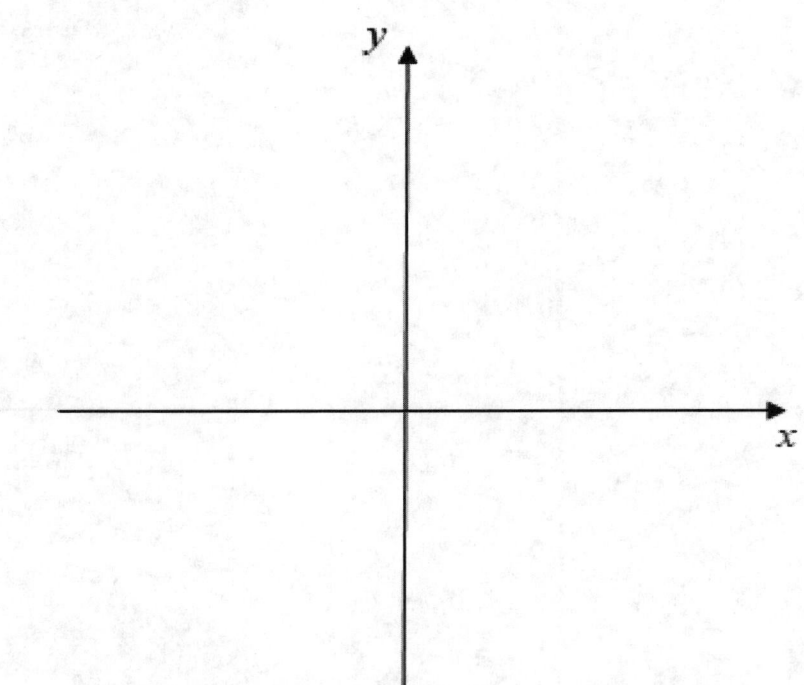

## Section 2.4 Guided Notebook

### 2.4 Inverse Trigonometric Functions I
- ☐ Work through Section 2.4 TTK #1–11
- ☐ Work through Section 2.4 Introduction
- ☐ Work through Section 2.4 Objective 1
- ☐ Work through Section 2.4 Objective 2
- ☐ Work through Section 2.4 Objective 3

### Section 2.4 Inverse Trigonometric Functions I

### 2.4 Things To Know

1. Determining Whether a Function is One-to-One Using the Horizontal Line Test
Try working through a "You Try It" problem or watch the video.

2. Understanding the Definition of an Inverse Function
Try working through a "You Try It" problem or watch the video.

3. Sketching the Graph of an Inverse Function
Try working through a "You Try It" problem or watch the animation.

4. Understanding the Special Right Triangles
Try working through a "You Try It" problem or watch the animation.

5. Understanding the Right Triangle Definitions of the Trigonometric Functions
Try working through a "You Try It" problem or watch the video.




6. Understanding the Signs of the Trigonometric Functions
Try working through a "You Try It" problem or watch the video.




7. Determining Reference Angles
Try working through a "You Try It" problem or watch the interactive video.




8. Evaluating Trigonometric Functions of Angles Belonging to the $\frac{\pi}{3}$, $\frac{\pi}{6}$, or $\frac{\pi}{4}$ Families
Try working through a "You Try It" problem or watch the interactive video.




9. Understanding the Graph of the Sine Function and Its Properties
Try working through a "You Try It" problem or watch the video.

10. Understanding the Graph of the Cosine Function and Its Properties
Try working through a "You Try It" problem or watch the video.




11. Understanding the Graph of the Tangent Function and Its Properties
Try working through a "You Try It" problem.




Section 2.4 Introduction

Explain why the function $f(x) = x^2 + 1$ is **not** one-to-one.




How can you restrict the domain of $f(x) = x^2 + 1$ so that it is a one-to-one function?




Watch the animation that accompanies the Introduction and sketch the graph in the animation that is a one-to-one function.

Section 2.4

### Section 2.4 Objective 1 Understanding and Finding the Exact and Approximate Values of the Inverse Sine Function

Read through this objective and watch the animation that describes the inverse sine function. Take your animation notes here:

What is the definition of the **Inverse Sine Function**?
(Define in words and sketch the graph.)

Section 2.4

What are the four **Steps for Determining the Exact Value of** $\sin^{-1}x$ ?

**Step 1.**

**Step 2.**

**Step 3.**

**Step 4.**

Section 2.4

Work through the interactive video with Example 1 showing all work below. Determine the exact value of each expression.

a. $\sin^{-1}\left(\dfrac{1}{2}\right)$

b. $\sin^{-1}\left(-\dfrac{\sqrt{3}}{2}\right)$

Section 2.4

Work through Example 2 showing all work below.
Use a calculator to approximate each value, or state that the value does not exist.

a. $\sin^{-1}(.7)$

b. $\sin^{-1}(-.95)$

c. $\sin^{-1}(3)$

Section 2.4

## Section 2.4 Objective 2 Understanding and Finding the Exact and Approximate Values of the Inverse Cosine Function

What is the definition of the **Inverse Cosine Function**? (Define and sketch.)

What are the four **Steps for Determining the Exact Value of $\cos^{-1} x$**?

**Step 1.**

**Step 2.**

**Step 3.**

**Step 4.**

Section 2.4

Work through the interactive video with Example 3 showing all work below.
Determine the exact value of each expression.

a. $\cos^{-1}(1)$

b. $\cos^{-1}\left(-\dfrac{1}{\sqrt{2}}\right)$

Work through Example 4 showing all work below.
Use a calculator to approximate each value, or state that the value does not exist.

a. $\cos^{-1}(1.5)$

b. $\cos^{-1}(-.25)$

134

Section 2.4

Section 2.4 Objective 3 Understanding and Finding the Exact and Approximate Values of the Inverse Tangent Function

What is the definition the **Inverse Tangent Function**? (Define and sketch.)

What are the four **Steps for Determining the Exact Value of** $\tan^{-1}x$?

**Step 1.**

**Step 2.**

**Step 3.**

**Step 4.**

Section 2.4

Work through the interactive video with Example 5 showing all work below.
Determine the exact value of each expression.

a. $\tan^{-1}\left(\dfrac{1}{\sqrt{3}}\right)$

b. $\tan^{-1}\left(-\dfrac{1}{\sqrt{3}}\right)$

Work through Example 6 showing all work below.
Use a calculator to approximate the value of $\tan^{-1}(20),$ or state that the value does not exist.

**Section 2.5 Guided Notebook**

## 2.5 Inverse Trigonometric Functions II
- ☐ Work through Section 2.5 TTK #1–9
- ☐ Work through Section 2.5 Objective 1
- ☐ Work through Section 2.5 Objective 2
- ☐ Work through Section 2.5 Objective 3
- ☐ Work through Section 2.5 Objective 4

## Section 2.5 Inverse Trigonometric Functions II

### 2.5 Things To Know

1. Understanding the Composition Cancellation Equations
Try working through a "You Try It" problem or watch the video.

2. Understanding the Special Right Triangles
Try working through a "You Try It" problem or watch the animation.

3. Understanding the Right Triangle Definitions of the Trigonometric Functions
Try working through a "You Try It" problem or watch the video.

4. Understanding the Signs of the Trigonometric Functions
Try working through a "You Try It" problem or watch the video.

Section 2.5

5. Determining Reference Angles
Try working through a "You Try It" problem or watch the interactive video.

6. Evaluating Trigonometric Functions of Angles Belonging to the $\frac{\pi}{3}$, $\frac{\pi}{6}$, or $\frac{\pi}{4}$ Families
Try working through a "You Try It" problem or watch the interactive video.

7. Understanding the Inverse Sine Function
Try working through a "You Try It" problem or watch the interactive video.

8. Understanding the Inverse Cosine Function
Try working through a "You Try It" problem or watch the interactive video.

9. Understanding the Inverse Tangent Function
Try working through a "You Try It" problem or watch the interactive video.

## Introduction to Section 2.5

Review the graphs of $y = \sin^{-1} x$, $y = \cos^{-1} x$, and $y = \tan^{-1} x$. Sketch these graphs below and state the domain and range of each.

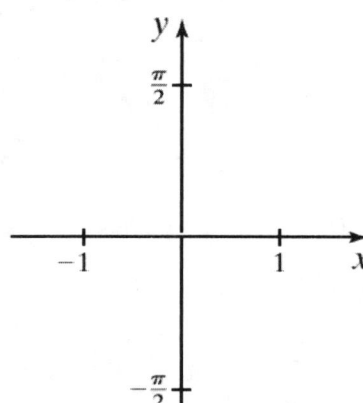

$y = \sin^{-1} x$

Domain:

Range:

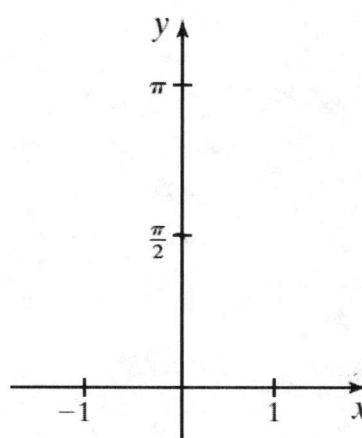

$y = \cos^{-1} x$

Domain:

Range:

$y = \tan^{-1} x$

Domain:

Range:

Section 2.5

## Section 2.5 Objective 1 Evaluating Composite Functions Involving Inverse Trigonometric Functions of the Form $f \circ f^{-1}$ and $f^{-1} \circ f$

What are the **Cancellation Equations for Compositions of Inverse Trigonometric Functions**?

Work through the interactive video with Example 1 showing all work below.
Find the exact value of each expression or state that it does not exist.

a. $\sin\left(\sin^{-1} \dfrac{1}{2}\right)$

b. $\cos\left(\cos^{-1} \dfrac{3}{2}\right)$

c. $\tan(\tan^{-1}(8.2))$

d. $\sin(\sin^{-1}(1.3))$

Section 2.5

Work through the interactive video with Example 2 showing all work below.
Find the exact value of each expression or state that it does not exist.

a. $\sin^{-1}\left(\sin\dfrac{\pi}{6}\right)$

b. $\cos^{-1}\left(\cos\dfrac{2\pi}{3}\right)$

c. $\sin^{-1}\left(\sin\dfrac{4\pi}{3}\right)$

d. $\tan^{-1}\left(\tan\dfrac{7\pi}{10}\right)$

Section 2.5

## Section 2.5 Objective 2 Evaluating Composite Functions Involving Inverse Trigonometric Functions of the Form $f \circ g^{-1}$ and $f^{-1} \circ g$

Work through the interactive video with Example 3 showing all work below.
Find the exact value of each expression or state that it does not exist.

a. $\cos(\tan^{-1}\sqrt{3})$

b. $\csc\left(\cos^{-1}\left(-\dfrac{\sqrt{3}}{2}\right)\right)$

c. $\sec\left(\sin^{-1}\left(-\dfrac{\sqrt{5}}{8}\right)\right)$

Work through the interactive video with Example 4 showing all work below.
Find the exact value of each expression or state that it does not exist.

a. $\sin^{-1}\left(\cos\left(-\dfrac{2\pi}{3}\right)\right)$

b. $\cos^{-1}\left(\sin\dfrac{\pi}{7}\right)$

Section 2.5 Objective 3 Understanding the Inverse Cosecant, Inverse Secant, and Inverse Cotangent Functions

What is the definition of the **Inverse Cosecant Function**? (Define and sketch.)

Section 2.5

What is the definition of the **Inverse Secant Function**? (Define and sketch.)

What is the definition of the **Inverse Cotangent Function**? (Define and sketch.)

Work through Example 5 showing all work below.
Find the exact value of $\sec^{-1}(-\sqrt{2})$ or state that it does not exist.

Section 2.5

Work through Example 6 showing all work below.
Use a calculator to approximate each value or state that the value does not exist.

a. $\sec^{-1}(5)$

b. $\cot^{-1}(-10)$

c. $\csc^{-1}(0.4)$

Section 2.5 Objective 4 Writing Trigonometric Expressions as Algebraic Expressions

Work through the video with Example 7 showing all work below.
Rewrite the trigonometric expression $\sin(\tan^{-1} u)$ as an algebraic expression involving the variable $u$. Assume that $\tan^{-1} u$ represents an angle whose terminal side is located in Quadrant I.

# Section 3.1 Guided Notebook

## 3.1 Trigonometric Identities
- [ ] Work through Section 3.1 TTK #1–3
- [ ] Work through Section 3.1 Objective 1
- [ ] Work through Section 3.1 Objective 2
- [ ] Work through Section 3.1 Objective 3
- [ ] Work through Section 3.1 Objective 4
- [ ] Work through Section 3.1 Objective 5
- [ ] Work through Section 3.1 Objective 6
- [ ] Work through Section 3.1 Objective 7
- [ ] Work through Section 3.1 Objective 8

## Section 3.1 Trigonometric Identities

### 3.1 Things To Know

1. Understanding the Quotient Identities for Acute Angles
Try working through a "You Try It" problem or watch the video.

2. Understanding the Reciprocal Identities for Acute Angles
Try working through a "You Try It" problem or watch the video.

3. Understanding the Pythagorean Identities for Acute Angles
Try working through a "You Try It" problem or watch the video.

Section 3.1

## Section 3.1 Objective 1 Reviewing the Fundamental Identities

What are the two **Quotient Identities**?

What are the six **Reciprocal Identities**?

What are the three **Pythagorean Identities**?

What are the four **Odd Properties**?

What are the two **Even Properties**?

## Section 3.1 Objective 2 Substituting Known Identities to Verify an Identity

Work through the interactive video with Example 1 showing all work below. Verify each identity.

a. $\tan x \cot x = 1$

b. $\sec^2 3x + \cot^2 3x - \tan^2 3x = \csc^2 3x$

c. $(5\sin y + 2\cos y)^2 + (5\cos y - 2\sin y)^2 = 29$

Section 3.1

## Section 3.1 Objective 3 Changing to Sines and Cosines to Verify an Identity

Work through the interactive video with Example 2 showing all work below. Verify each identity.

a. $\sin^2 t = \tan t \cot t - \cos^2 t$

b. $\dfrac{\sec\theta \csc\theta}{\cot\theta} = \sec^2\theta$

c. $\dfrac{\cos(-\theta)}{\sec\theta} + \sin(-\theta)\csc\theta = -\sin^2\theta$

Section 3.1 Objective 4 Factoring to Verify an Identity

Write down the following special factoring formulas:

**Difference of Two Squares:**

**Perfect Square Formulas:**

**Sum of Two Cubes:**

**Difference of Two Cubes:**

Section 3.1

Work through the interactive video with Example 3 and show all work below.
Verify each trigonometric identity.

a. $\sin x - \cos^2 x \sin x = \sin^3 x$

b. $\dfrac{\tan^3 \alpha - 1}{\tan \alpha - 1} = \sec^2 \alpha + \tan \alpha$

c. $\dfrac{8\sin^2 \theta - 2\sin \theta - 3}{1 + 2\sin \theta} = 4\sin \theta - 3$

## Section 3.1 Objective 5 Separating a Single Quotient into Multiple Quotients to Verify an Identity

When one side of a trigonometric identity is a quotient of the form $\dfrac{A+B}{C}$, where $C$ is a single trigonometric expression, then we can separate the expression into two expressions using the property:

$$\frac{A+B}{C} =$$

(Fill in the right side of the equals sign above.)

Work through the video with Example 4 and show all work below.

Verify the trigonometric identity $\dfrac{\sin\alpha + \cos\alpha}{\cos\alpha} - \dfrac{\sin\alpha + \cos\alpha}{\sin\alpha} = \tan\alpha - \cot\alpha$.

Section 3.1

## Section 3.1 Objective 6 Combining Fractional Expressions to Verify an Identity

Work through the video with Example 5 and show all work below.

Verify the trigonometric identity $\dfrac{1-\csc\theta}{\cot\theta} - \dfrac{\cot\theta}{1-\csc\theta} = 2\tan\theta$.

Section 3.1

## Section 3.1 Objective 7 Multiplying by Conjugates to Verify Identities

Given the expression $A + B$, then the conjugate is _____.

Work through the video with Example 6 and show all work below.
Verify the trigonometric identity $\dfrac{\sin\theta}{\csc\theta + 1} = \dfrac{1 - \sin\theta}{\cot^2\theta}$.

Section 3.1

Section 3.1  Objective 8 Summarizing the Techniques for Verifying Identities

What are the six techniques described in **A Summary for Verifying Trigonometric Identities**?

1.

2.

3.

4.

5.

6.

Work through the interactive video with Example 7 and show all work below.
Verify each trigonometric identity.

a. $\dfrac{\sin^2 t + 6\sin t + 9}{\sin t + 3} = \dfrac{3\csc t + 1}{\csc t}$

Section 3.1

b. $\dfrac{2\csc\theta}{\sec\theta} + \dfrac{\cos\theta}{\sin\theta} = 3\cot\theta$

c. $\dfrac{1-\sin\theta}{\cos\theta} + \dfrac{\cos\theta}{1-\sin\theta} = 2\sec\theta$

## Section 3.2 Guided Notebook

### 3.2 The Sum and Difference Formulas
- [ ] Work through Section 3.2 TTK #1–4
- [ ] Work through Section 3.2 Objective 1
- [ ] Work through Section 3.2 Objective 2
- [ ] Work through Section 3.2 Objective 3
- [ ] Work through Section 3.2 Objective 4
- [ ] Work through Section 3.2 Objective 5

### Section 3.2 The Sum and Difference Formulas

#### 3.2 Things To Know

1. Understanding Cofunctions
Try working through a "You Try It" problem or watch the video.

2. Evaluating Trigonometric Functions of Angles Belonging to the $\frac{\pi}{3}$, $\frac{\pi}{6}$, or $\frac{\pi}{4}$ Families
Try working through a "You Try It" problem or watch the interactive video.

3. Finding the Exact and Approximate Values of an Inverse Sine Expression
Try working through a "You Try It" problem or watch the interactive video.

4. Finding the Exact and Approximate Values of an Inverse Cosine Expression
Try working through a "You Try It" problem or watch the interactive video.

Section 3.2

Section 3.2 Objective 1 Understanding the Sum and Difference Formulas for the Cosine Function

What are the **Sum and Difference Formulas for the Cosine Function**?

Work through the interactive video with Example 1 showing all work below.
Find the exact value of each trigonometric expression without the use of a calculator.

a. $\cos\left(\dfrac{2\pi}{3} + \dfrac{3\pi}{4}\right)$

b. $\cos(225° - 150°)$

Work through the interactive video with Example 2 showing all work below.
Find the exact value of each trigonometric expression without the use of a calculator.

a. $\cos\left(\dfrac{7\pi}{12}\right)\cos\left(\dfrac{5\pi}{12}\right) + \sin\left(\dfrac{7\pi}{12}\right)\sin\left(\dfrac{5\pi}{12}\right)$

b. $\cos\left(\dfrac{7\pi}{12}\right)$

c. $\cos(-75°)$

Section 3.2

Work through the video with Example 3 showing all work below.
Suppose that the terminal side of angle $\alpha$ lies in Quadrant IV and the terminal side of angle $\beta$ lies in Quadrant III. If $\cos\alpha = \dfrac{4}{7}$ and $\sin\beta = -\dfrac{8}{13}$, find the exact value of $\cos(\alpha+\beta)$.

## Section 3.2 Objective 2 Understanding the Sum and Difference Formulas for the Sine Function

What are the two **Cofunction Identities for Sine and Cosine**?

What are the **Sum and Difference Formulas for the Sine Function**?

Work through the interactive video with Example 4 showing all work below.
Find the exact value of each trigonometric expression without the use of a calculator.

a. $\sin\left(-\dfrac{\pi}{3} + \dfrac{5\pi}{4}\right)$

Section 3.2

b. $\sin(12°)\cos(78°) + \cos(12°)\sin(78°)$

c. $\sin(15°)$

Work through the video with Example 5 showing all work below.
Suppose that $\alpha$ is an angle such that $\tan\alpha = \dfrac{5}{7}$ and $\cos\alpha < 0$. Also, suppose that $\beta$ is an angle such that $\sec\beta = -\dfrac{4}{3}$ and $\csc\beta > 0$. Find the exact value of $\sin(\alpha+\beta)$.

## Section 3.2 Objective 3 Understanding the Sum and Difference Formulas for the Tangent Function

What are the two **Sum and Difference Formulas for the Tangent Function?**




Work through the interactive video with Example 6 showing all work below.
Find the exact value of each trigonometric expression without the use of a calculator.

a. $\tan\left(\dfrac{5\pi}{6} + \dfrac{3\pi}{4}\right)$




b. $\tan\left(\dfrac{\pi}{12}\right)$

## Section 3.2 Objective 4 Using the Sum and Difference Formulas to Verify Identities

Work through Example 7 and show all work below.
Verify the trigonometric identity $\sin(2\theta) = 2\sin\theta\cos\theta$.

Work through the video with Example 8 and show all work below.
Verify the trigonometric identity $\csc(\alpha - \beta) = \dfrac{\sin\alpha\cos\beta + \cos\alpha\sin\beta}{\sin^2\alpha - \sin^2\beta}$.

## Section 3.2 Objective 5 Using the Sum and Difference Formulas to Evaluate Expressions Involving Inverse Trigonometric Functions

Work through the video with Example 9 and show all work below.

Find the exact value of the expression $\cos\left(\sin^{-1}\left(\dfrac{1}{5}\right) + \cos^{-1}\left(-\dfrac{3}{4}\right)\right)$ without using a calculator.

## Section 3.3 Guided Notebook

### 3.3 The Double-Angle and Half-Angle Formulas
- [ ] Work through Section 3.3 TTK #1–6
- [ ] Work through Section 3.3 Objective 1
- [ ] Work through Section 3.3 Objective 2
- [ ] Work through Section 3.3 Objective 3
- [ ] Work through Section 3.3 Objective 4
- [ ] Work through Section 3.3 Objective 5

### Section 3.3 The Double-Angle and Half-Angle Formulas

#### 3.3 Things To Know

1. Evaluating Trigonometric Functions of Angles Belonging to the $\frac{\pi}{3}, \frac{\pi}{6},$ or $\frac{\pi}{4}$ Families
Try working through a "You Try It" problem or watch the interactive video.

2. Finding the Exact and Approximate Values of an Inverse Sine Expression
Try working through a "You Try It" problem or watch the interactive video.

3. Finding the Exact and Approximate Values of an Inverse Cosine Expression
Try working through a "You Try It" problem or watch the interactive video.

Section 3.3

4. Finding the Exact and Approximate Values of an Inverse Tangent Expression
Try working through a "You Try It" problem or watch the interactive video.

5. Understanding the Sum and Difference Formulas for the Sine Function
Try working through a "You Try It" problem or watch the interactive video.

6. Understanding the Sum and Difference Formulas for the Cosine Function
Try working through a "You Try It" problem or watch the interactive video.

Section 3.3  Objective 1 Understanding the Double-Angle Formulas

What are the three **Double-Angle Formulas**?

What are the three **Double-Angle Formulas for Cosine**?

Section 3.3

Work through the interactive video with Example 1 showing all work below.
Rewrite each expression as the sine, cosine, or tangent of a double angle.
Then evaluate the expression without using a calculator.

a. $\cos^2\left(\dfrac{11\pi}{12}\right) - \sin^2\left(\dfrac{11\pi}{12}\right)$

b. $2\sin 67.5° \cos 67.5°$

c. $\dfrac{2\tan\left(-\dfrac{\pi}{8}\right)}{1 - \tan^2\left(-\dfrac{\pi}{8}\right)}$

d. $2\cos^2 105° - 1$

Section 3.3

Work through the interactive video with Example 2 showing all work below.

Suppose that the terminal side of an angle $\theta$ lies in Quadrant II such that $\sin\theta = \dfrac{5}{7}$. Find the values of $\sin 2\theta$, $\cos 2\theta$, and $\tan 2\theta$.

Section 3.3

Work through the video with Example 3 showing all work below.

If $\tan 2\theta = -\dfrac{24}{7}$ for $\dfrac{3\pi}{2} < 2\theta < 2\pi$, then find the values of $\sin\theta$, $\cos\theta$, and $\tan\theta$.

Section 3.3 Objective 2 Understanding the Power Reduction Formulas

What are the three **Power Reduction Formulas**?

Section 3.3

Work through the video with Example 4 showing all work below.
Rewrite the function $f(x) = 6\sin^4 x$ as an equivalent function containing only cosine terms raised to a power of 1.

Section 3.3  Objective 3 Understanding the Half-Angle Formulas

What are the **Half-Angle Formulas for Sine and Cosine**?

**What are the Half-Angle Formulas for Tangent?**

Work through Example 5 showing all work below.
Use a half-angle formulas to evaluate each expression without using a calculator.

a. $\sin\left(-\dfrac{7\pi}{8}\right)$

Section 3.3

b. cos(−15°)

c. $\tan\left(\dfrac{11\pi}{12}\right)$

Work through the interactive video with Example 6 showing all work below.
Suppose that $\csc\alpha = \dfrac{8}{3}$ such that $\dfrac{\pi}{2} < \alpha < \pi$. Find the values of $\sin\left(\dfrac{\alpha}{2}\right)$, $\cos\left(\dfrac{\alpha}{2}\right)$, and $\tan\left(\dfrac{\alpha}{2}\right)$.

Section 3.3

## Section 3.3 Objective 4 Using the Double-Angle, Power Reduction, and Half-Angle Formulas to Verify Identities

Work through Example 7 and show all work below.

Verify the trigonometric identity $\dfrac{\cos(2\theta)}{1+\sin(2\theta)} = \dfrac{\cos\theta - \sin\theta}{\cos\theta + \sin\theta}$.

## Section 3.3 Objective 5 Using the Double-Angle and Half-Angle Formulas to Evaluate Expressions Involving Inverse Trigonometric Functions

Work through the video with Example 8 and show all work below.

Find the exact value of the expression $\cos\left(\frac{1}{2}\sin^{-1}\left(-\frac{3}{11}\right)\right)$ without the use of a calculator.

## Section 3.4 Guided Notebook

### 3.4 The Product-to-Sum and Sum-to-Product Formulas
- ☐ Work through Section 3.3 TTK #1–3
- ☐ Work through Section 3.3 Objective 1
- ☐ Work through Section 3.3 Objective 2
- ☐ Work through Section 3.3 Objective 3

### Section 3.4 The Product-to-Sum and Sum-to-Product Formulas

### 3.4 Things To Know

1. Evaluating Trigonometric Functions of Angles Belonging to the $\frac{\pi}{3}$, $\frac{\pi}{6}$, or $\frac{\pi}{4}$ Families

Try working through a "You Try It" problem or watch the interactive video.

2. Understanding the Sum and Difference Formulas for the Sine Function
Try working through a "You Try It" problem or watch the interactive video.

3. Understanding the Sum and Difference Formulas for the Cosine Function
Try working through a "You Try It" problem or watch the interactive video.

Section 3.4

Section 3.4 Objective 1 Understanding the Product-to-Sum Formulas

What are the four **Product-to-Sum Formulas**?

Work through the interactive video with Example 1 showing all work below.
Write each product as a sum or difference containing only sines or cosines.

a. $\sin 4\theta \sin 2\theta$

b. $\cos\left(\dfrac{19\theta}{2}\right)\sin\left(\dfrac{\theta}{2}\right)$

c. $\cos 11\theta \cos 5\theta$

d. $\sin 6\theta \cos 3\theta$

Section 3.4

Work through the video with Example 2 showing all work below. Determine the exact value of the expression $\sin\left(\frac{3\pi}{8}\right)\cos\left(\frac{\pi}{8}\right)$ without the use of a calculator.

Section 3.4 Objective 2 Understanding the Sum-to-Product Formulas

What are the four **Sum-to-Product Formulas**?

Section 3.4

Work through the interactive video with Example 3 showing all work below.
Write each sum or difference as a product of sines and/or cosines.

a. $\sin 5\theta + \sin 3\theta$

b. $\cos\left(\dfrac{3\theta}{2}\right) - \cos\left(\dfrac{17\theta}{2}\right)$

Section 3.4

Work through the interactive video with Example 4 showing all work below. Determine the exact value of the expression $\sin\left(\dfrac{\pi}{12}\right) - \sin\left(\dfrac{17\pi}{12}\right)$ without the use of a calculator.

## Section 3.4 Objective 3 Using the Product-to-Sum and Sum-to-Product Formulas to Verify Identities

Work through Example 5 showing all work below.

Verify the trigonometric identity $\dfrac{\cos\theta + \cos 3\theta}{2\cos 2\theta} = \cos\theta$.

## Section 3.5 Guided Notebook

**3.5 Trigonometric Equations**
- ☐ Work through Section 3.5 TTK #1–4
- ☐ Work through Section 3.5 Objective 1
- ☐ Work through Section 3.5 Objective 2
- ☐ Work through Section 3.5 Objective 3
- ☐ Work through Section 3.5 Objective 4
- ☐ Work through Section 3.5 Objective 5

## Section 3.5 Trigonometric Equations

### 3.5 Things To Know

1. Solving Equations That Are Quadratic in Form
Try working through a "You Try It" problem or watch the interactive video.

2. Evaluating Trigonometric Functions of Angles Belonging to the $\frac{\pi}{3}$, $\frac{\pi}{6}$, or $\frac{\pi}{4}$ Families
Try working through a "You Try It" problem or watch the interactive video.

3. Finding the Exact and Approximate Values of an Inverse Sine Expression
Try working through a "You Try It" problem or watch the interactive video.

4. Finding the Exact and Approximate Values of an Inverse Cosine Expression
Try working through a "You Try It" problem or watch the interactive video.

Section 3.5

Section 3.5  Introduction

What is the difference between **identities** and **conditional trigonometric equations**?

What is the difference between **general solution(s)** and **specific solution(s)**?

**Watch the animation seen in the intro.**
First, sketch the functions $y = \sin\theta$ and $y = b$ for $-1 \le b \le 1$ and show at least 10 solutions to the equation $\sin\theta = b$.

**Continue to watch the animation seen in the intro.**
How many solutions are there to the equation $\sin\theta = b$ for $b > 1$?
Draw a picture and explain your answer.

How many solutions are there to the equation $\sin\theta = b$ for $b < -1$?
Draw a picture and explain your answer.

How many solutions are there to the equation $\sin\theta = b$ for $-1 \leq b \leq 1$ on the interval $0 \leq \theta < 2\pi$? Draw a picture and explain your answer.

Section 3.5

Section 3.5  Objective 1 Solving Trigonometric Equations That Are Linear in Form

What is a trigonometric equation that is linear in form?

Give at least 3 examples of trigonometric equations that are linear in form:

What are the four **Steps for Solving Trigonometric Equations That Are Linear in Form**?

**Step 1.**

**Step 2.**

**Step 3.**

**Step 4.**

Work through the interactive video with Example 1 showing all work below. Determine a general formula (or formulas) for all solutions to each equation. Then, determine the specific solutions (if any) on the interval $[0, 2\pi)$.

a. $\sin \theta = \dfrac{1}{2}$

b. $\sqrt{3} \tan \theta + 1 = 0$

Section 3.5

c. $\sec\theta = -1$

Work through the interactive video with Example 2 showing all work below. Determine a general formula (or formulas) for all solutions to each equation. Then, determine the specific solutions (if any) on the interval $[0, 2\pi)$.

a. $\sqrt{2}\cos 2\theta + 1 = 0$

b. $\sin\dfrac{\theta}{2} = -\dfrac{\sqrt{3}}{2}$

c. $\tan\left(\theta + \dfrac{\pi}{6}\right) + 1 = 0$

Section 3.5

Section 3.5 Objective 2 Solving Trigonometric Equations That Are Quadratic in Form

What is a trigonometric equation that is quadratic in form?

Work through the interactive video with Example 3 showing all work below. Determine a general formula (or formulas) for all solutions to each equation. Then, determine the specific solutions (if any) on the interval $[0, 2\pi)$.

a. $\sin^2 \theta - 4\sin \theta + 3 = 0$

b. $4\cos^2\theta - 3 = 0$

Section 3.5

## Section 3.5 Objective 3 Solving Trigonometric Equations Using Identities

Work through the interactive video with Example 4 showing all work below. Determine a general formula (or formulas) for all solutions to each equation. Then, determine the specific solutions (if any) on the interval $[0, 2\pi)$.

a. $2\sin^2\theta = 3\cos\theta + 3$

b. $\sin\theta\cos\theta = -\dfrac{1}{2}$

c. $\cos 2\theta + 4\sin^2 \theta = 2$

d. $\sin 5\theta + \sin 3\theta = 0$

## Section 3.5 Objective 4 Solving Other Types of Trigonometric Equations

Work through the interactive video with Example 5 and show all work below. Determine a general formula (or formulas) for all solutions to each equation. Then, determine the specific solutions (if any) on the interval $[0, 2\pi)$.

a. $\sin 2\theta + 2\cos\theta \sin 2\theta = 0$

b. $\cos^2 \theta = \sin\theta \cos\theta$

c. $\sin\theta + \cos\theta = 1$

Before we can attempt to multiply or divide both sides of a trigonometric equation by a function, what must we first verify? (Hint: Look in the Caution text that follows the solution to Example 5b.)

Section 3.5

## Section 3.5 Objective 5 Solving Trigonometric Equations Using a Calculator

Work through Example 6 and show all work below.
Approximate all solutions to the trigonometric equation $\cos\theta = -0.4$ on the interval $[0, 2\pi)$.
Round your answer to four decimal places.

Why can't we simply use an inverse trigonometric function key on our calculator to solve this trigonometric equation? (Hint: See the Caution text that follows Example 6.)

## Section 4.1 Guided Notebook

### 4.1 Right Triangle Applications
- Work through Section 4.1 TTK #1 and 2
- Work through Section 4.1 Objective 1
- Work through Section 4.1 Objective 2

## Section 4.1 Right Triangle Applications

### 4.1 Things To Know

1. Using the Pythagorean Theorem
Try working through a "You Try It" problem or watch the video.

2. Evaluating Trigonometric Functions Using a Calculator
Try working through a "You Try It" problem or watch the video.

### Section 4.1 Objective 1 Solving Right Triangles

Fill in the blanks below:

The goal when solving a right triangle is to determine the _____ of all angles

and the length of _____ given certain information.

Section 4.1

Work through Example 1 showing all work below.
Solve the given right triangle. Round the length of side $a$ and the measure of the two acute angles to two decimal places.

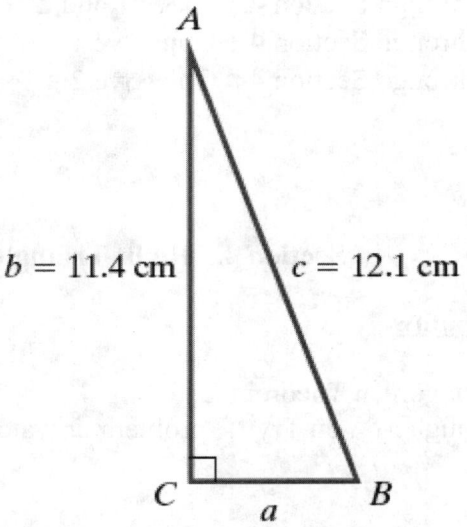

Work through the video with Example 2 showing all work below.
Suppose that the length of the hypotenuse of a right triangle is 11 inches. If one of the acute angles is $37.34°$, find the length of the two legs and the measure of the other acute angle. Round all values to two decimal places.

Section 4.1

Try solving the right triangles below on your own:

1. $a = 4.9$, $b =$ _____, $c =$ _____
   $A =$ _____, $B = 27°$

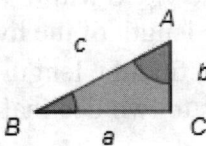

2. $a =$ _____, $b =$ _____, $c = 44.6$
   $A = 67°$, $B =$ _____

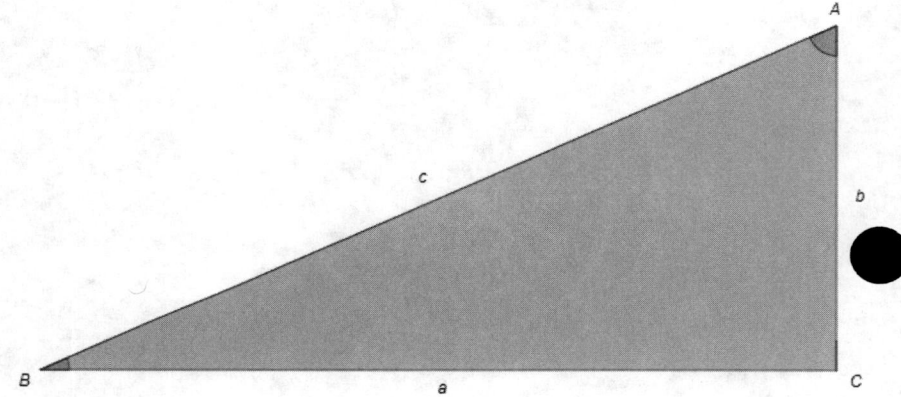

To check your answers above, open the **Guided Visualization** seen on page 4.1-11 or your eText. Input the parameters for the two triangles above to check your solutions. You may want to come up with a couple more triangles to solve to make sure that you understand this concept.

206

Section 4.1 Objective 2 Applications of Right Triangle Trigonometry

What is the **angle of elevation**?

What is the **angle of depression**?

Sketch and label the diagram seen on the first page of this objective.

Section 4.1

Work through the video with Example 3 and show all work below.
The angle of elevation to the top of a flagpole measured by a digital protractor is $20°$ from a point on the ground 90 feet away from its base. Find the height of the flagpole. Round to two decimal places.

Work through the video with Example 4 and show all work below.
At the same instant, two observers 5 miles apart are looking at the same airplane. The angle of elevation (from the ground) of the observer closest to the plane is 71°. The angle of elevation (from the ground) of the person furthest from the plane is 39°. Find the altitude of the plane (to the nearest foot) at the instant the two people observe the plane.

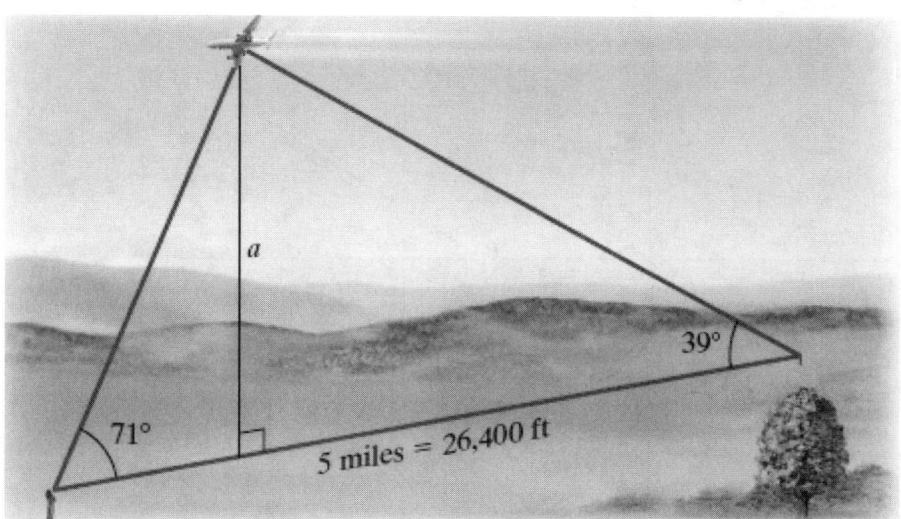

Section 4.1

Work through the video with Example 5 and show all work below.

A tourist visiting Paris determines that the angle of elevation from a point $A$ to the top of the Eiffel tower is $12.19°$. She then walks 1 km on a straight line toward the tower to point $B$ and determines that the angle of elevation is to the top of the tower is $32.94°$. Determine the height of the Eiffel Tower. Round to the nearest meter.

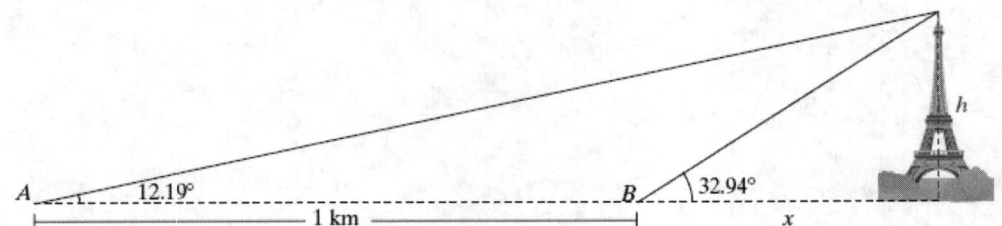

**Section 4.2 Guided Notebook**

**4.2 The Law of Sines**
- Work through Section 4.2 TTK #1
- Work through Section 4.2 Objective 1
- Work through Section 4.2 Objective 2
- Work through Section 4.2 Objective 3
- Work through Section 4.2 Objective 4

**Section 4.2 The Law of Sines**

## 4.2 Things To Know

1. Understanding the Inverse Sine Function
Try working through a "You Try It" problem or watch the interactive video.

## Section 4.2 Objective 1 Determining If the Law of Sines Can be Used to Solve an Oblique Triangle

What are the two types of **oblique triangles**? (Describe and sketch.)

Section 4.2

What is the **Law of Sines**?

Describe the six cases of oblique triangles, as seen in Table 1.

Section 4.2

What three pieces of information are needed to solve an oblique triangle using the Law of Sines?

1.

2.

3.

Work through the video with Example 1 showing all work below.
Decide whether or not the Law of Sines can be used to solve each triangle.
Do not attempt to solve the triangle.

a.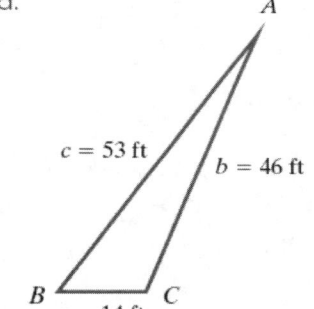

b.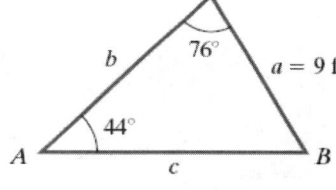

c.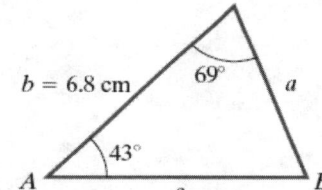

Section 4.2

## Section 4.2 Objective 2 Using the Law of Sines to Solve the SAA Case or ASA Case

Work through Example 2 showing all work below.
Solve the given oblique triangle. Round the lengths of the sides to one decimal place.

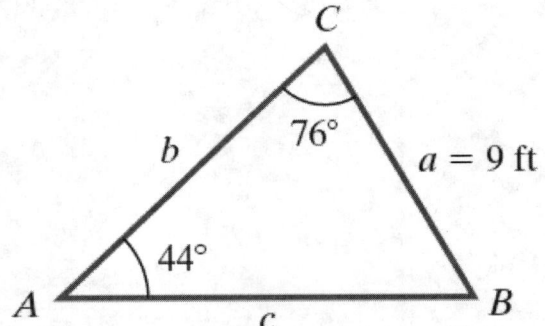

Work through the video with Example 3 showing all work below.
Solve the oblique triangle $ABC$ if $B = 38°$, $C = 72°$, and $a = 7.5$ cm.
Round the lengths of the sides to one decimal place.

Section 4.2

Section 4.2 Objective 3 Using the Law of Sines to Solve the SSA Case

Watch the animation found in this objective to see all of the possible triangles that can result when given the SSA case and take notes here:

Describe the different scenarios for the SSA case as seen in Table 2.

Work through Example 4 and show all work below.
Two sides and an angle are given below. Determine whether the information results in no triangle, one right triangle, or one or two oblique triangles. Solve each resulting triangle. Round the measures of all angles and the lengths of all sides to one decimal place.

$a = 10$ ft,  $b = 28$ ft,  $A = 29°$

Section 4.2

Work through the video with Example 5 and show all work below.
Two sides and an angle are given below. Determine whether the information results in no triangle, one right triangle, or one or two oblique triangles. Solve each resulting triangle. Round the measures of all angles and the lengths of all sides to one decimal place.

$a = 13$ cm,  $b = 7.8$ cm,  $A = 67°$

Section 4.2

- Work through the video with Example 6 and show all work below.
Two sides and an angle are given below. Determine whether the information results in no triangle, one right triangle, or one or two oblique triangles. Solve each resulting triangle. Round the measures of all angles and the lengths of all sides to one decimal place.

$b = 11.3$ in., $c = 15.5$ in., $B = 34.7°$

- What are the three cases for which the Law of Sines can be used? (Sketch and label.)

Section 4.2

## Section 4.2 Objective 4 Using the Law of Sines to Solve Applied Problems Involving Oblique Triangles

Work through Example 7 and show all work below.
To determine the width of a river, forestry workers place markers on opposite sides of the river at points A and B. A third marker is placed at point C, 200 feet away from point A. If the angle between point C and B is 51° and if the angle between point A and point C is 110°, then determine the width of the river rounded to the nearest tenth of a foot.

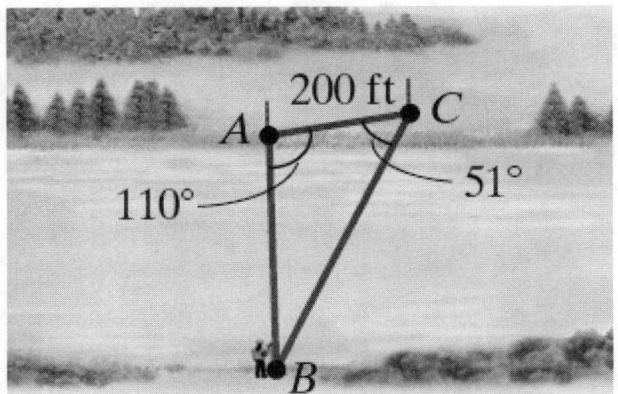

Section 4.2

- Describe the concept of **bearing**.



- Work through the video with Example 8 and show all work below.
A ship set sail from port at a bearing of N 53° E and sailed 63 km to point $B$. The ship then turned and sailed an additional 69 km to point $C$. Determine the distance from port to point $C$ if the ship's final bearing is N 74° E. Round to the nearest tenth of a kilometer.

# Section 4.3 Guided Notebook

## 4.3 The Law of Cosines
- ☐ Work through Section 4.3 TTK #1 and 2
- ☐ Work through Section 4.3 Objective 1
- ☐ Work through Section 4.3 Objective 2
- ☐ Work through Section 4.3 Objective 3
- ☐ Work through Section 4.3 Objective 4

## Section 4.3 The Law of Cosines

### 4.3 Things To Know

1. Understanding the Inverse Sine Function
Try working through a "You Try It" problem or watch the animation and/or interactive video.




2. Understanding the Inverse Cosine Function
Try working through a "You Try It" problem or watch the animation and/or interactive video.

Section 4.3

## Section 4.3 Objective 1 Determining If the Law of Sines or the Law of Cosines Should Be Used to Begin to Solve an Oblique Triangle

The Law of Sines can be used to solve which three cases of triangles?

What is the **Law of Cosines**?

What is the **Alternate Form of the Law of Cosines**?

Section 4.3

Work through the video with Example 1 showing all work below.
Decide whether the Law of Sines or the Law of Cosines should be used to begin to solve the given triangle. Do not solve the triangle.

a. Triangle with $b = 7.8$ cm, angle $C = 79.5°$, $c = 13.9$ cm, side $a$ opposite $A$.

b. Triangle with angle $C = 72°$, angle $A = 78°$, angle $B = 38°$, sides $a$, $b$, $c$.

c. Triangle with $a = 5$, $c = 3$, angle $B = 98°$, side $b$.

Section 4.3 Objective 2 Using the Law of Cosines to Solve the SAS Case

What are the three steps for **Solving an SAS Oblique Triangle**?

**Step 1.**

**Step 2.**

**Step 3.**

Section 4.3

Work through the interactive video with Example 2 showing all work below.
Solve the given oblique triangle. Round the measures of all angles and the lengths of all sides to one decimal place.

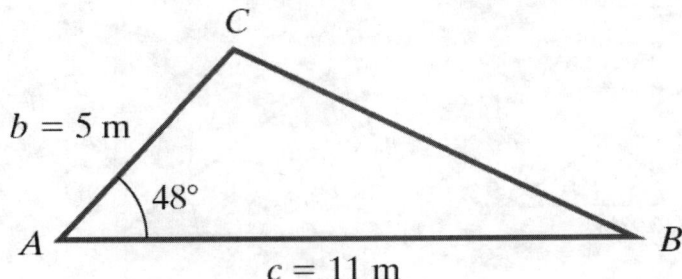

Section 4.3  Objective 3 Using the Law of Cosines to Solve the SSS Case

What are the three steps for **Solving a SSS Oblique Triangle**?

**Step 1.**

**Step 2.**

**Step 3.**

Work through the interactive video with Example 3 and show all work below.
Solve oblique triangle $ABC$ if $a = 5$ ft, $b = 8$ ft, and $c = 12$ ft.

Section 4.3

## Section 4.3 Objective 4 Using the Law of Cosines to Solve Applied Problems Involving Oblique Triangles

Work through the video with Example 4 and show all work below.

Two planes take off from different runways at the same time. One plane flies at an average speed of 350 mph with a bearing of N 21° E. The other plane flies at an average speed of 420 mph with a bearing of S 84° W. How far are the planes from each other 2 hours after takeoff? Round to the nearest tenth of a mile.

Work through Example 5 and show all work below.
A **chord** of a circle is a line segment with endpoints that both lie on the circumference of a circle. Determine the measure of the central angle (in degrees) if the length of the chord intercepted by the central angle of a circle of radius 10 inches is 16.5 inches. Round the measure of the central angle to one decimal place.

## Section 4.4 Guided Notebook

**4.4 Area of Triangles**
- ☐ Work through Section 4.4 TTK #1–4
- ☐ Work through Section 4.4 Objective 1
- ☐ Work through Section 4.4 Objective 2
- ☐ Work through Section 4.4 Objective 3

## Section 4.4 Area of Triangles

### 4.4 Things To Know

1. Understanding the Inverse Sine Function
Try working through a "You Try It" problem or watch the interactive video.

2. Understanding the Inverse Cosine Function
Try working through a "You Try It" problem or watch the interactive video.

3. Using the Law of Sines to Solve the SAA Case or the ASA Case
Try working through a "You Try It" problem or watch the video.

4. Using the Law of Cosines to Solve the SSS Case
Try working through a "You Try It" problem or watch the interactive video.

Section 4.4

Section 4.4 Objective 1 Determining the Area of Oblique Triangles

How do we compute the **Area of a Triangle** in terms of the length of the base and the height?

If $A$, $B$, and $C$ are the measures of the angles of any triangle and if $a$, $b$, and $c$ are the lengths of the sides opposite the corresponding angles, then the area of triangle $ABC$ is given by what three formulas?

1. Area =

2. Area =

3. Area =

Work through the video with Example 1 showing all work below.
Determine the area of each triangle. Round each answer to two decimal places.

a.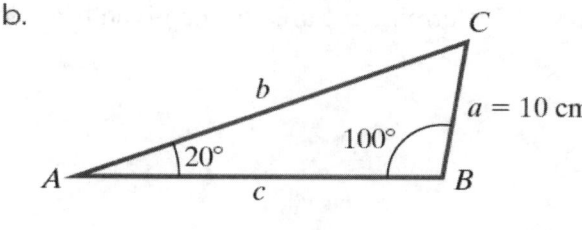

b.

Section 4.4

## Section 4.4 Objective 2 Using Heron's Formula to Determine the Area of an SSS Triangle

Work through Example 2 showing all work below.
Determine the area of the given triangle. Round to two decimal places.

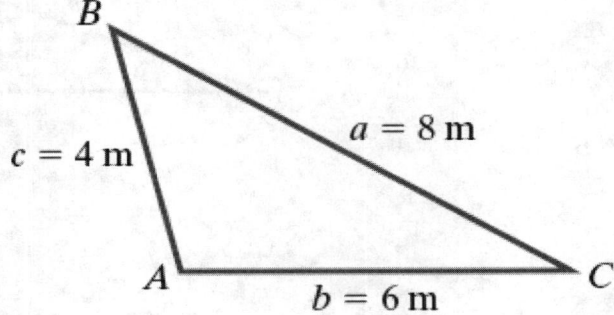

Section 4.4

What is **Heron's formula**?

Work through the video with Example 3 showing all work below.
Use Heron's formula to determine the area of the given triangle.
Round to two decimal places.

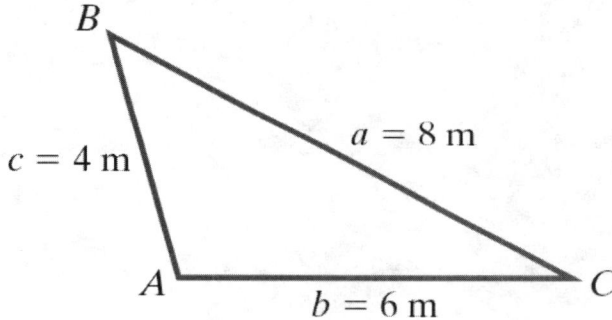

Section 4.4

## Section 4.4  Objective 3 Solving Applied Problems Involving the Area of Triangles

Work through Example 4 and show all work below.
A painter who is painting a house has only one side of the house left to paint. He has enough paint to cover 1200 square feet. A cross section of the unpainted side of the house is shown in the figure. What is the area of the unpainted side of the house? Does he have enough paint to finish the job?

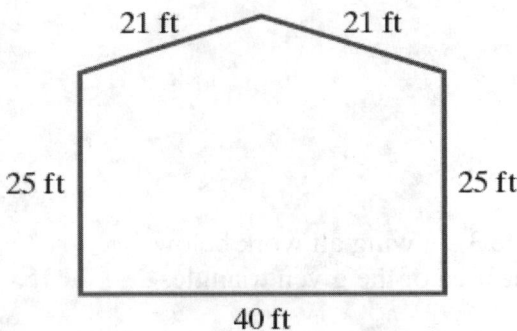

## Section 5.1 Guided Notebook

### 5.1 Polar Coordinates and Polar Equations
- ☐ Work through Section 5.1 TTK #1–5
- ☐ Work through Section 5.1 Objective 1
- ☐ Work through Section 5.1 Objective 2
- ☐ Work through Section 5.1 Objective 3
- ☐ Work through Section 5.1 Objective 4
- ☐ Work through Section 5.1 Objective 5
- ☐ Work through Section 5.1 Objective 6

### Section 5.1 Polar Coordinates and Polar Equations

### 5.1 Things To Know

1. Understanding the Four Families of Special Angles
Try working through a "You Try It" problem or watch the interactive video.

2. Understanding the Definitions of the Trigonometric Functions of General Angles
Try working through a "You Try It" problem or watch the video.

3. Understanding the Signs of the Trigonometric Functions
Try working through a "You Try It" problem or watch the video.

Section 5.1

4. Evaluating Trigonometric Functions of Angles Belonging to the $\frac{\pi}{3}$, $\frac{\pi}{6}$, or $\frac{\pi}{4}$ Families

Try working through a "You Try It" problem or watch the interactive video.

5. Solving Trigonometric Equations That Are Linear in Form

Try working through a "You Try It" problem or watch the interactive video.

Section 5.1  Objective 1 Plotting Points Using Polar Coordinates

What is the **polar coordinate system**?

What is the definition of **Directed Distance**?

Section 5.1

Sketch and label a polar grid.

Work through the video with Example 1 showing all work below.
Plot the following points in a polar coordinate system.

a. $A\left(3, \dfrac{\pi}{4}\right)$

b. $B(-2, 120°)$

c. $C\left(1.5, -\dfrac{7\pi}{6}\right)$

d. $D\left(-3, -\dfrac{3\pi}{4}\right)$

Section 5.1

Click on the **Guided Visualization** titled "Plotting Polar Coordinates," which can be found on the bottom of page 5.1-9.

After opening this **Guided Visualization,** plot polar coordinates having the following parameters.

1. Plot and label a point in the polar coordinate system such that $r > 0$ and $\theta < 0$.

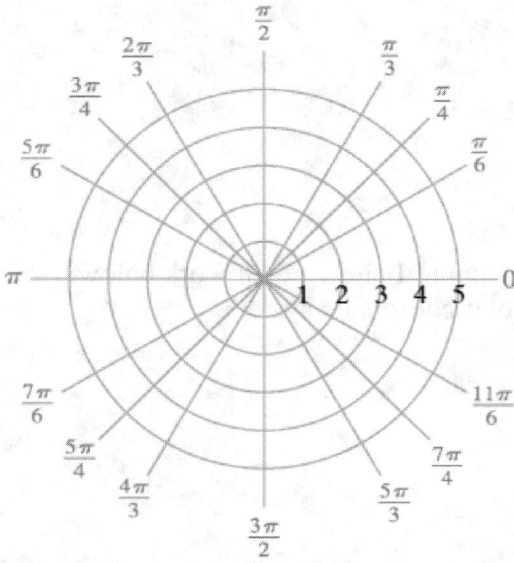

2. Plot and label a point in the polar coordinate system such that $r < 0$ and $\theta > 0$.

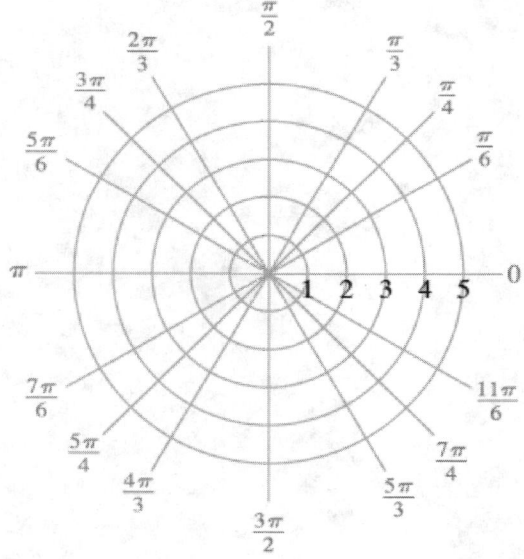

240

Copyright © 2019 Pearson Education, Inc.

Section 5.1 Objective 2 Determining Different Representations of the Point (r, θ)

In the video that accompanies this objective (this is the same video as seen in Example 1), the points $A\left(3, \dfrac{\pi}{4}\right)$ and $D\left(-3, -\dfrac{3\pi}{4}\right)$ are positioned in the exact same location. Plot these two points on the same polar grid below to convince yourself that these two points are, indeed, positioned at the same location.

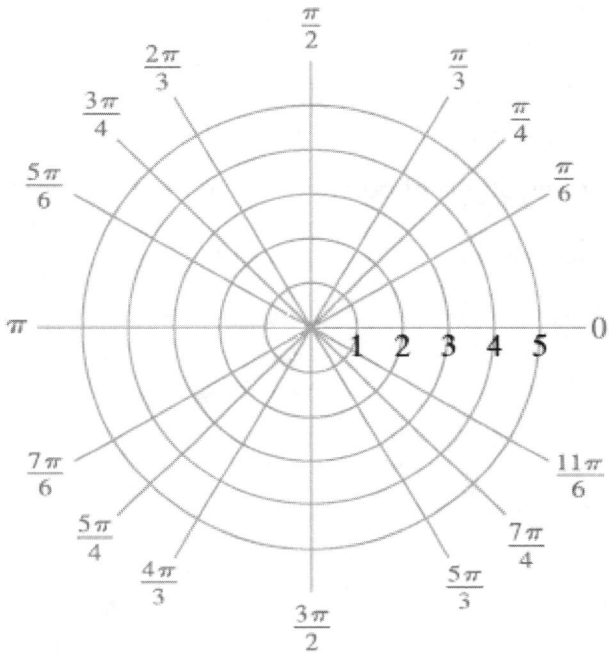

Write down the two ways to determine different representations for a point $(r, \theta)$.

Section 5.1

Work through the video with Example 2 showing all work below.

The point $P\left(4, \dfrac{5\pi}{6}\right)$ is shown below:

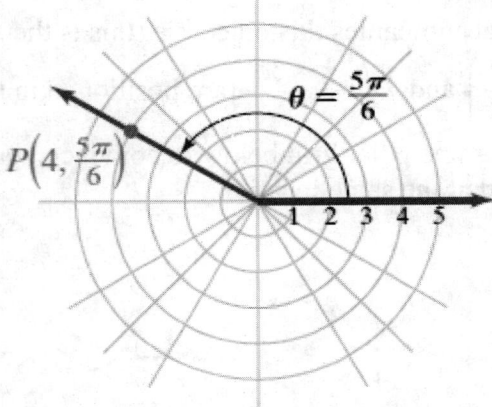

Determine three different representations of point $P$ that have the specified conditions.

a. $r > 0,\ -2\pi \leq \theta < 0$

b. $r < 0,\ 0 \leq \theta < 2\pi$

c. $r > 0,\ 2\pi \leq \theta < 4\pi$

Section 5.1 Objective 3 Converting a Point from Polar Coordinates to Rectangular Coordinates

What are the **Relationships Used when Converting a Point from Polar Coordinates to Rectangular Coordinates**?

Work through the video with Example 3 and show all work below.
Determine the rectangular coordinates for the points with the given polar coordinates.

a. $A(5, \pi)$

b. $B\left(-7, -\dfrac{\pi}{3}\right)$

c. $C\left(3\sqrt{2}, \dfrac{5\pi}{4}\right)$

Section 5.1

### Section 5.1 Objective 4 Converting a Point from Rectangular Coordinates to Polar Coordinates

Fill in the blanks below:
For simplicity and consistency, we will always determine the polar coordinates with the conditions that _____ and _____.


Work through the video with Example 4 and show all work below.
Determine the polar coordinates for the points with the given rectangular coordinates.

a. $A(-3.5, 0)$




b. $B(0, -\sqrt{7})$

Section 5.1

What are the four steps for **Converting Rectangular Coordinates to Polar Coordinates for Points Not Lying Along an Axis**?

**Step 1.**

**Step 2.**

**Step 3.**

**Step 4.**

Section 5.1

Work through the interactive video with Example 5 and show all work below.
Determine the polar coordinates for the points with the given rectangular coordinates such that $r \geq 0$ and $0 \leq \theta < 2\pi$.

a. $A(-4,-4)$

b. $B(-2\sqrt{3}, 2)$

c. $C(4,-3)$

Section 5.1

## Section 5.1 Objective 5 Converting an Equation from Rectangular Form to Polar Form

What is a **polar equation**? (Define and give examples.)

Work through the video with Example 6 and show all work below.
Convert each equation given in rectangular form into polar form.

a. $x = 7$

b. $2x - y = 8$

c. $4x^2 + 4y^2 = 3$

d. $x^2 + y^2 = 9y$

Section 5.1

## Section 5.1 Objective 6 Converting an Equation from Polar Form to Rectangular Form

Work through the video with Example 7 and show all work below.
Convert each equation given in polar form into rectangular form.

a. $3r\cos\theta - 4r\sin\theta = -1$

b. $r = 6\cos\theta$

c. $r = 3$

d. $\theta = \dfrac{\pi}{6}$

**Section 5.2 Guided Notebook**

**5.2 Graphing Polar Equations**
- [ ] Work through Section 5.2 TTK #1–4
- [ ] Work through Section 5.2 Objective 1
- [ ] Work through Section 5.2 Objective 2
- [ ] Work through Section 5.2 Objective 3
- [ ] Work through Section 5.2 Objective 4
- [ ] Work through Section 5.2 Objective 5

**Section 5.2 Graphing Polar Equations**

**5.2 Things To Know**

1. Evaluating Trigonometric Functions of Angles Belonging to the $\frac{\pi}{3}$, $\frac{\pi}{6}$, or $\frac{\pi}{4}$ Families
Try working through a "You Try It" problem or watch the interactive video.

2. Solving Trigonometric Equations That Are Linear in Form
Try working through a "You Try It" problem or watch the interactive video.

Section 5.2

3. Plotting Points Using Polar Coordinates
Try working through a "You Try It" problem or watch the video.

4. Converting an Equation from Polar Form to Rectangular Form
Try working through a "You Try It" problem or watch the video.

Section 5.2  Objective 1 Sketching Equations of the Form $\theta = a$, $r\cos\theta = a$, $r\sin\theta = a$, and $ar\cos\theta + br\sin\theta = c$

Work through the video with Example 1 showing all work below.

Sketch the graph of the polar equation $\theta = \dfrac{2\pi}{3}$.

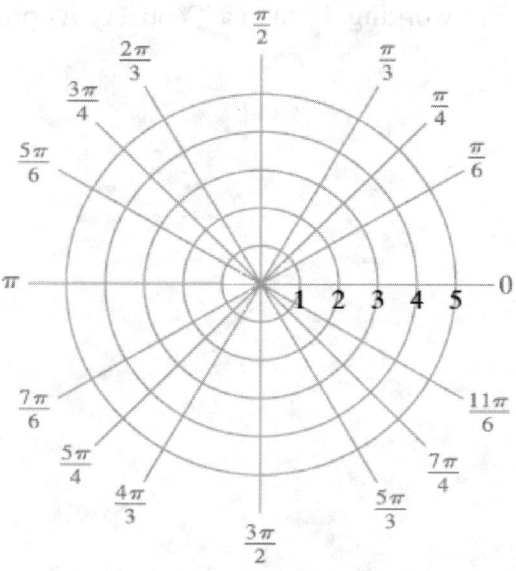

250

Copyright © 2019 Pearson Education, Inc.

- Work through Example 2 showing all work below.
  Sketch the graph of the polar equation $r\cos\theta = 2$.

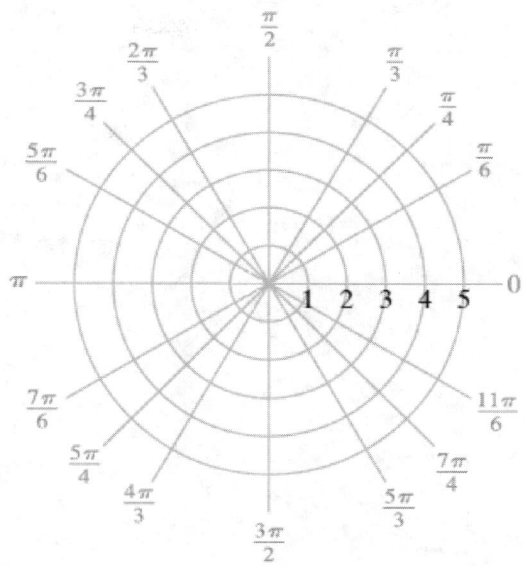

- Work through the video with Example 3 showing all work below.
  Sketch the graph of the polar equation $r\sin\theta = -3$.

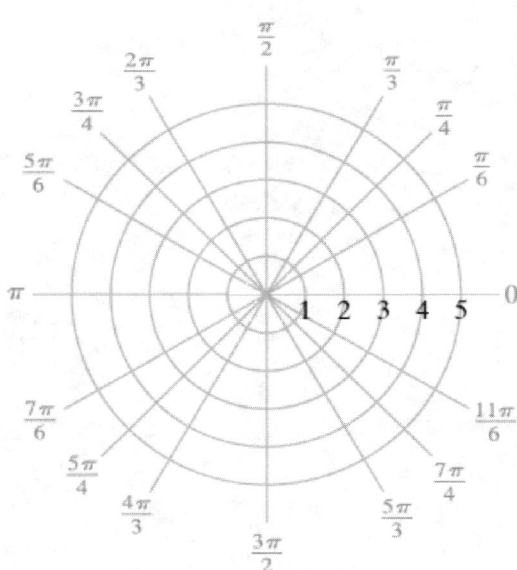

Section 5.2

Work through Example 4 showing all work below.
Sketch the graph of the polar equation $3r\cos\theta - 2r\sin\theta = -6$

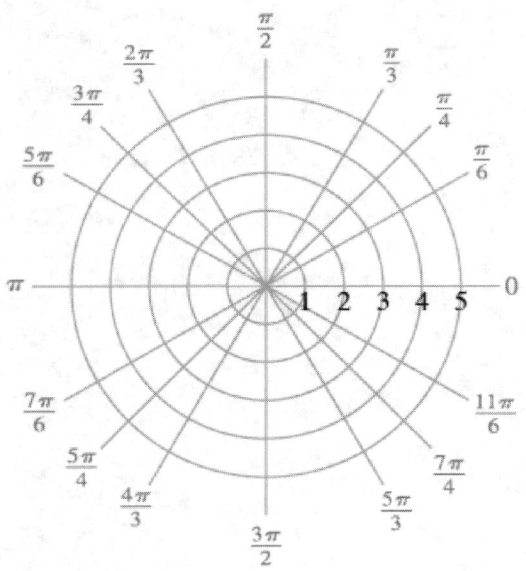

- Write down the summary, **Graphs of Polar Equations of the Form $\theta = \alpha$, $r\cos\theta = a$, $r\sin\theta = a$, and $ar\cos\theta + br\sin\theta = c$, where $a$, $b$, and $c$ are Constants.** (Be sure to sketch each graph.)

## Section 5.2 Objective 2 Sketching Equations of the Form $r = a$, $r = a\sin\theta = a$, and $r = a\cos\theta$

Write the polar equation $r = a$ in rectangular form.

Work through Example 5 showing all work below.
Sketch the graph of the polar equation $r = -3$.

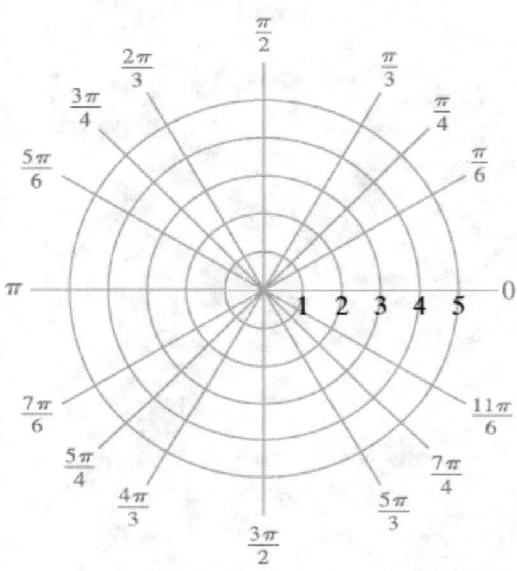

Work through the interactive video with Example 6 showing all work below.
Sketch the graph of each polar equation.

a. $r = 4\sin\theta$

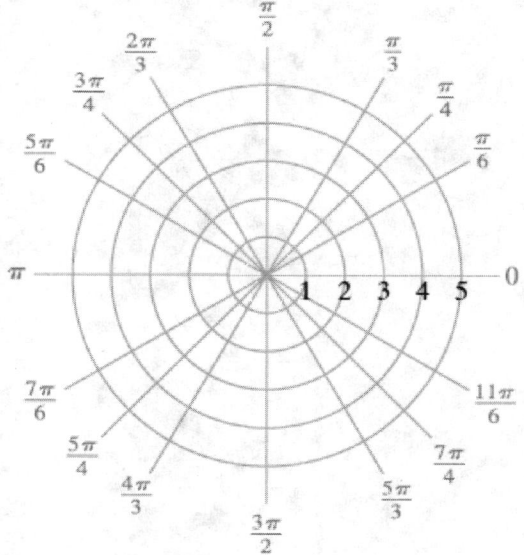

b. $r = -2\cos\theta$

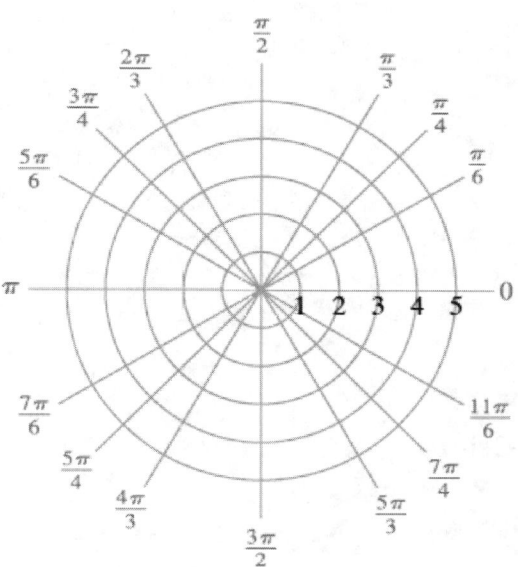

Section 5.2

Click on the **Guided Visualization** on the bottom of page 5.2.17. Use this **Guided Visualization** to sketch the following functions.

1. $r = 2$

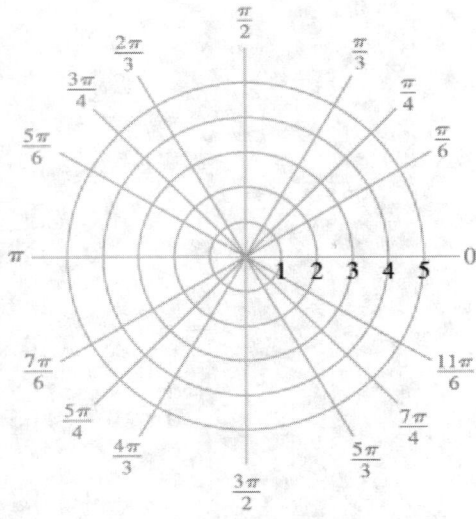

2. $r = -3\sin\theta$

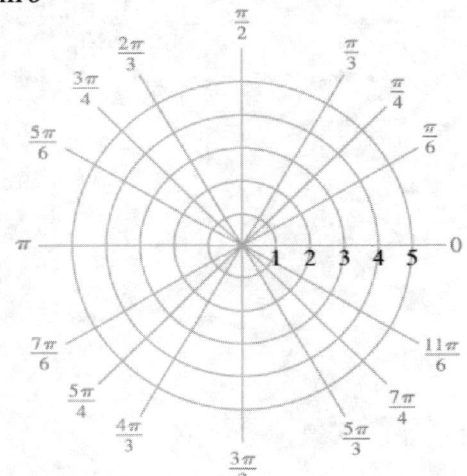

3. $r = 2\cos\theta$

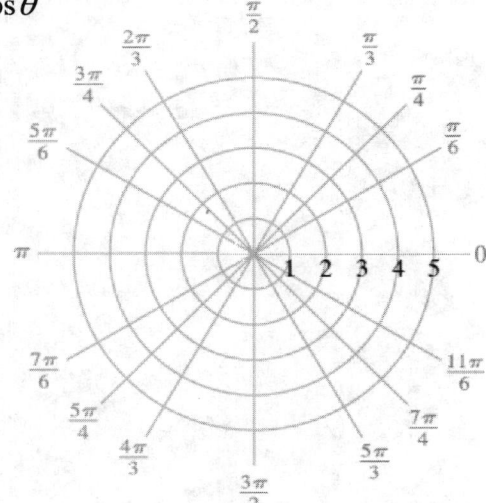

Write down the summary, **Graphs of Polar Equations of the Form $r = a$, $r = a\sin\theta$, and $r = a\cos\theta$, where $a \neq 0$ is a Constant.** (Be sure to sketch the graph of each.)

Section 5.2

## Section 5.2 Objective 3 Sketching Equations of the Form $r = a + b\sin\theta$ and $r = a + b\cos\theta$

What are **limacons**? What form do their equations have?

Work through Example 7 showing all work below.
Sketch the graph of the polar equation $r = 3 - 3\sin\theta$.

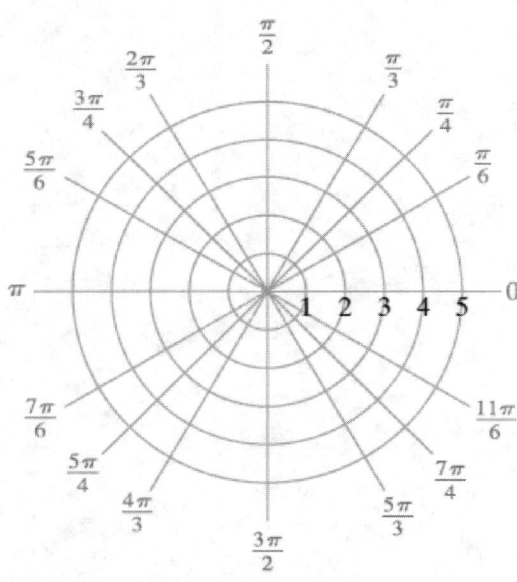

Section 5.2

What is a **cardioid**?

Work through Example 8 showing all work below.
Sketch the graph of the polar equation $r = -1 + 2\cos\theta$.

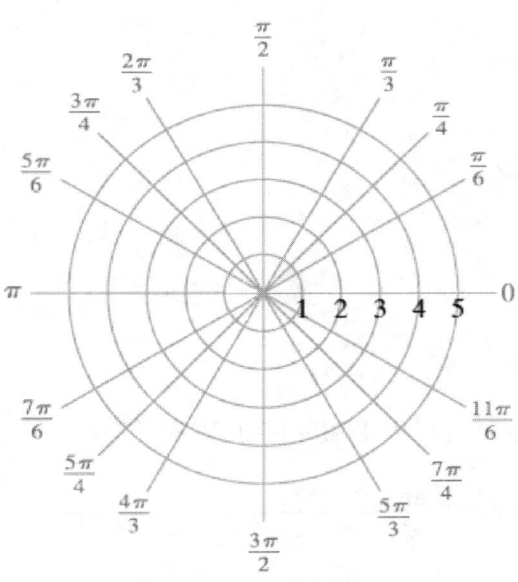

The graph from Example 8 is an example of a limacon that has

an _____.

259

Section 5.2

Work through Example 9 showing all work below.
Sketch the graph of the polar equation $r = 3 + 2\sin\theta$.

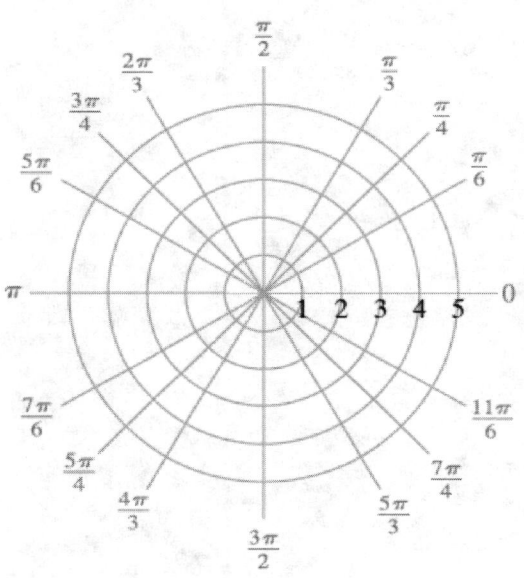

The graph from Example 9 is an example of a limacon that has

an_____.

260

Write down the summary, **Graphs of Polar Equations of the Form** $r = a + b\sin\theta$, **and** $r = a + b\cos\theta$, **where** $a \neq 0$ **and** $b \neq 0$ **Are Constants.** (Be sure to sketch **one graph** of each type.)

Section 5.2

What are the four **Steps for Sketching Polar Equations (Limacons) of the Form** $r = a + b\sin\theta$ and $r = a + b\cos\theta$ ?

**Step 1.**

**Step 2.**

**Step 3.**

**Step 4.**

Click on the **Guided Visualization** on page 5.2.31. See if you can first sketch the graph of $r = 1 + 2\cos\theta$ using the four-step process outlined on the previous page. Then, use this **Guided Visualization** on page 5.2-31 to check your answer.

Sketch the graph of $r = 1 + 2\cos\theta$.

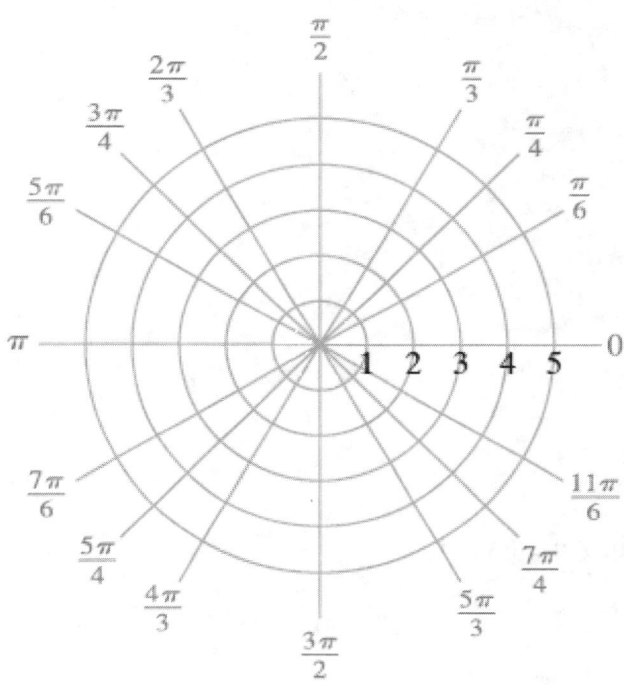

Section 5.2

Work through the interactive video with Example 10 showing all work below.
Sketch the graph of each polar equation.

a. $r = 4 - 3\cos\theta$

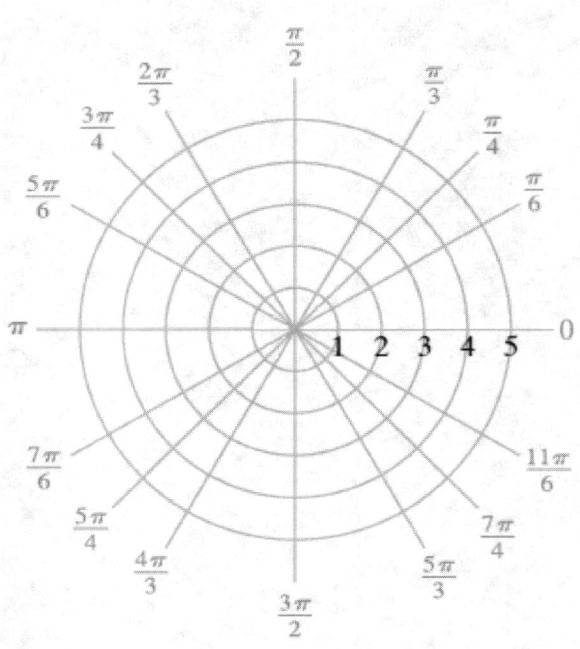

b. $r = 2 + \sin\theta$

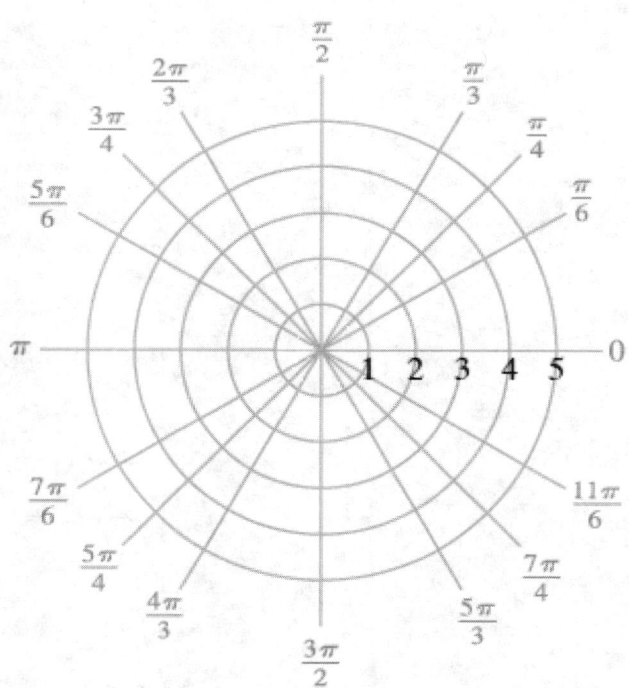

c. $r = -2 + 2\cos\theta$

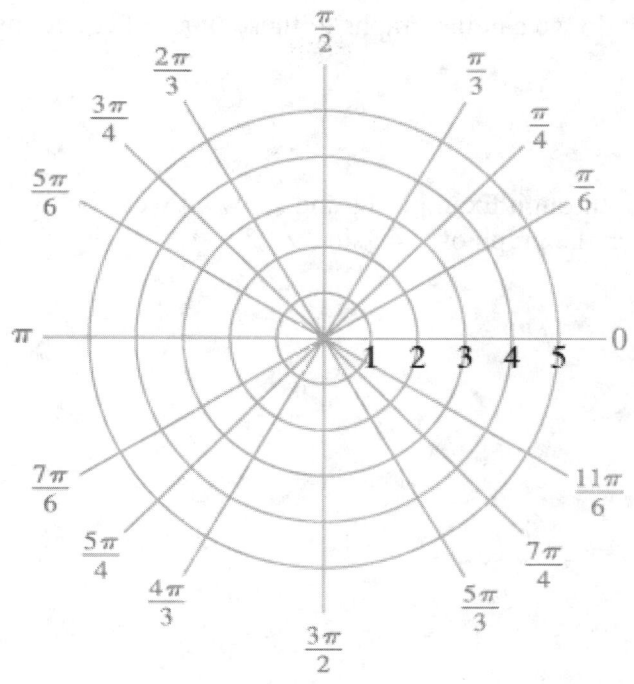

d. $r = 3 - 4\sin\theta$

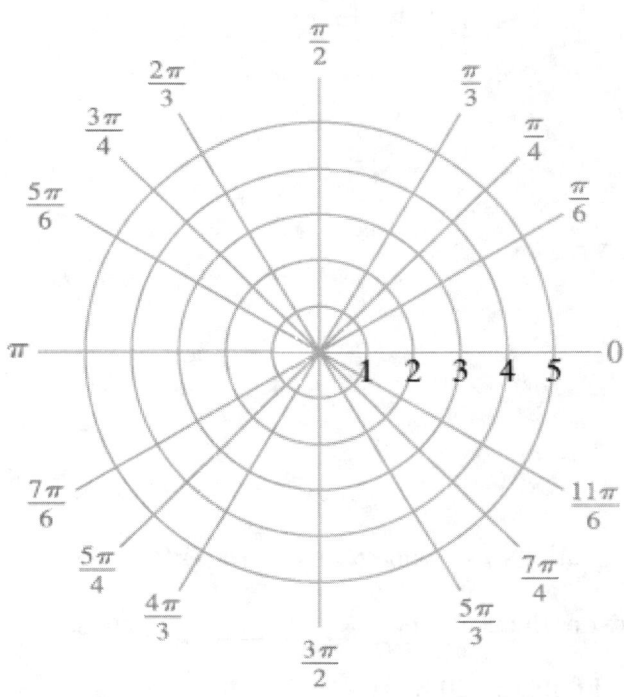

Section 5.2

## Section 5.2 Objective 4 Sketching Equations of the Form $r = a \sin n\theta$ and $r = a \cos n\theta$

What do we call the graphs of these types of equations? Why?

Work through Example 11 and show all work below.
Sketch the graph of $r = 3\sin 2\theta$.

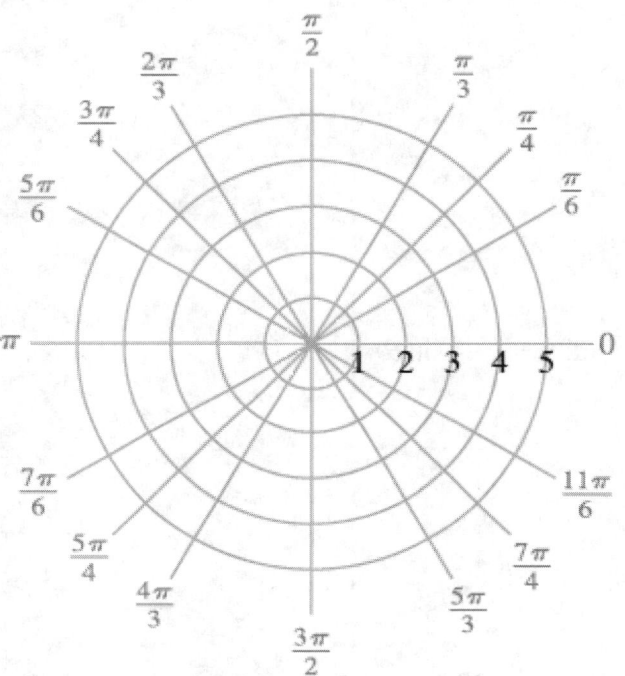

In the equations $r = a\sin n\theta$ or $r = a\cos n\theta$,

if $n$ is **even**, there will be _____ petals.

if $n$ is **odd**, there will be _____ petals.

Section 5.2

Write down the summary, **Graphs of Polar Equations of the Form $r = a \sin n\theta$, and $r = a \cos n\theta$, where $a \neq 0$ Is a Constant and $n \neq 1$ Is a Positive Integer.**
(Be sure to sketch a graph of each type.)

Section 5.2

What are the six **Steps for Sketching Polar Equations (Roses) of the Form** $r = a \sin n\theta$, and $r = a \cos n\theta$, where $a \neq 0$ **Is a Constant and** $n \neq 1$ **Is a Positive Integer?**

**Step 1.**

**Step 2.**

**Step 3.**

**Step 4.**

**Step 5.**

**Step 6.**

Click on the **Guided Visualization** on the bottom of page 5.2.40. See if you can first sketch the graph of $r = -2\sin 3\theta$ using the six-step process outlined on the previous page. Then, use this **Guided Visualization** on page 5.2-40 to check your answer.

Sketch the graph of $r = -2\sin 3\theta$.

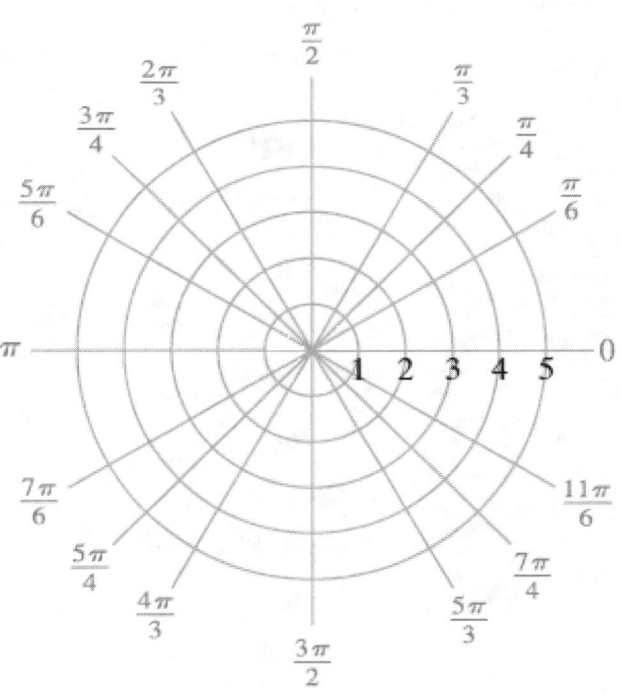

Section 5.2

Work through the interactive video with Example 12 showing all work below.
Sketch the graph of each polar equation.

a. $r = -4\cos 3\theta$

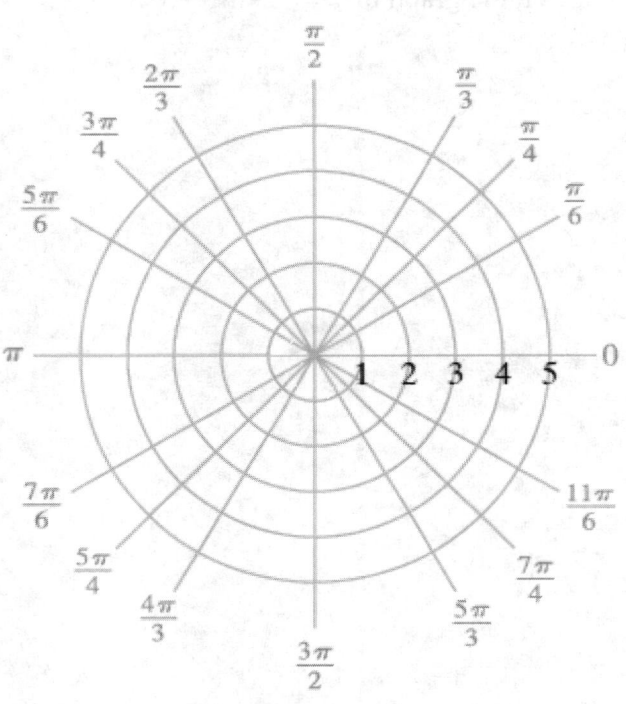

b. $r = -2\sin 5\theta$

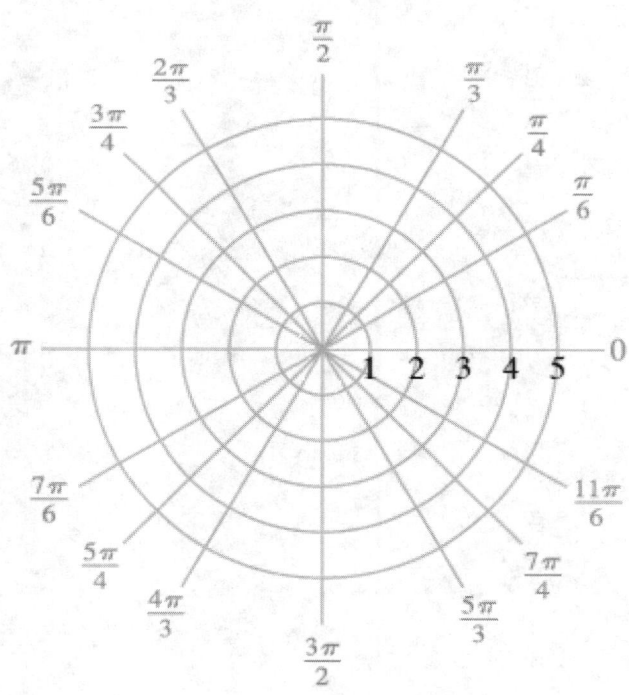

c. $r = 5\cos 4\theta$

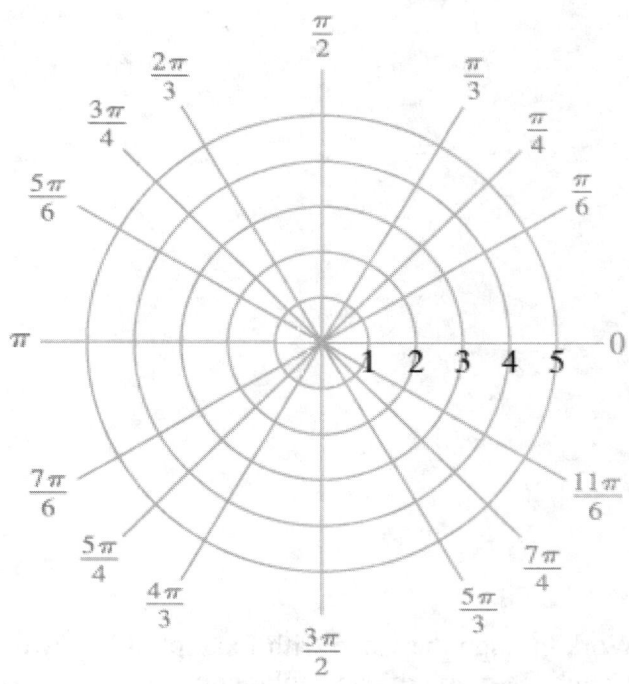

Section 5.2 Objective 5 Sketching Equations of the Form $r^2 = a^2 \sin 2\theta$ and $r^2 = a^2 \cos 2\theta$

What is a **lemniscate**? What form do their equations have?

Section 5.2

Write down the summary, **Graphs of Polar Equations of the Form $r^2 = a^2 \sin 2\theta$ and $r^2 = a^2 \cos 2\theta$, where $a \neq 0$ Is a Constant.**
(Be sure to sketch a graph of each type.)

Work through the video with Example 13 showing all work below.
Sketch the graph of each polar equation.

a. $r^2 = 9\cos 2\theta$

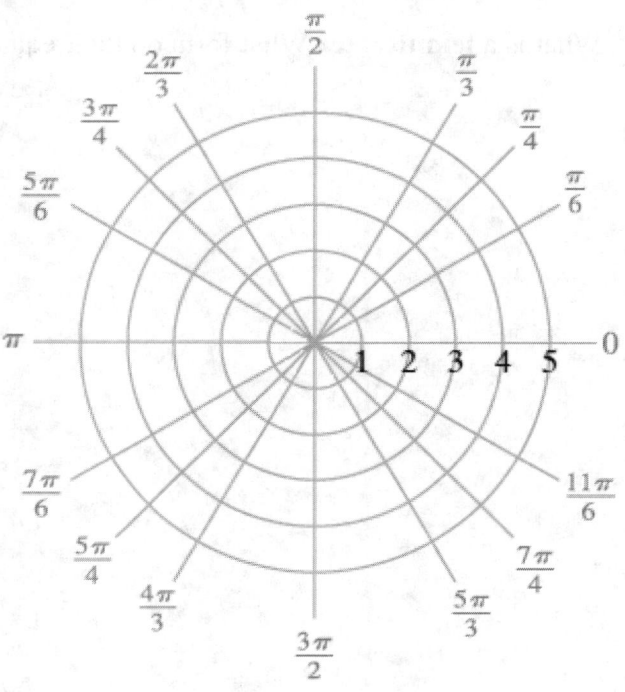

b. $r^2 = 16\sin 2\theta$

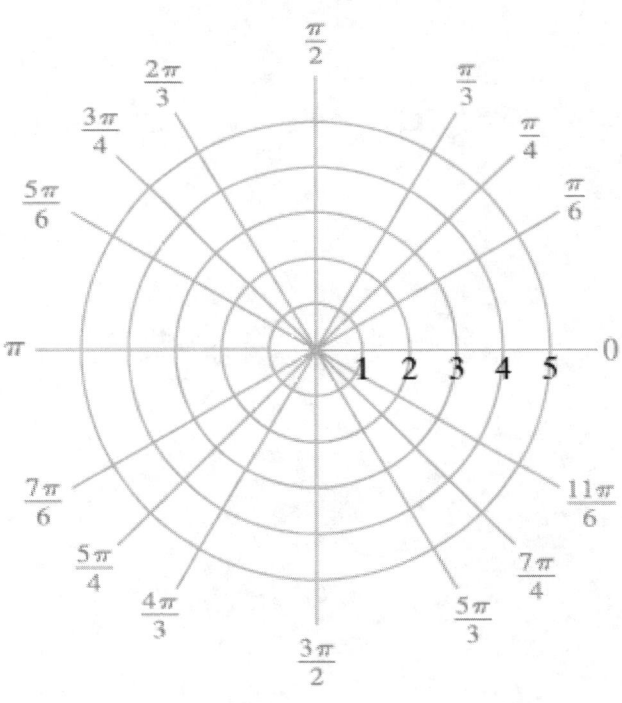

## Section 5.3 Guided Notebook

**5.3 Complex Numbers in Polar Form; De Moivre's Theorem**
- ☐ Work through Section 5.3 TTK #3–6
- ☐ Work through Section 5.3 Objective 1
- ☐ Work through Section 5.3 Objective 2
- ☐ Work through Section 5.3 Objective 3
- ☐ Work through Section 5.3 Objective 4
- ☐ Work through Section 5.3 Objective 5
- ☐ Work through Section 5.3 Objective 6
- ☐ Work through Section 5.3 Objective 7

**Section 5.3 Complex Numbers in Polar Form; De Moivre's Theorem**

**5.3 Things To Know**

3. Evaluating Trigonometric Functions of Angles Belonging to the $\frac{\pi}{3}$, $\frac{\pi}{6}$, or $\frac{\pi}{4}$ Families
Try working through a "You Try It" problem or watch the interactive video.

4. Solving Trigonometric Equations That Are Linear in Form
Try working through a "You Try It" problem or watch the interactive video.

5. Plotting Points Using Polar Coordinates
Try working through a "You Try It" problem or watch the video.

6. Converting a Point from Rectangular Coordinates to Polar Coordinates
Try working through a "You Try It" problem or watch the interactive video.

Section 5.3

Section 5.3  Objective 1 Understanding the Rectangular Form of a Complex Number

What is the definition of the **Rectangular Form of a Complex Number**?

What is the **complex plane**? (Define and sketch.)

What is the definition of the **Absolute Value of a Complex Number**?

Section 5.3

Work through the video with Example 1 showing all work below.
Plot each complex number in the complex plane and determine its absolute value.

a. $z_1 = 3 - 4i$

b. $z_2 = -2 + 5i$

c. $z_3 = 3$

d. $z_4 = -2i$

Section 5.3

## Section 5.3 Objective 2 Understanding the Polar Form of a Complex Number

What is the definition of the **Polar Form of a Complex Number**?

What is the **modulus**? What is the **argument**?

Complex numbers in polar form, where $0 \leq \theta < 2\pi$ (or $0 \leq \theta < 360°$) are said to be in _____.

Section 5.3

Work through the video with Example 2 showing all work below.
Rewrite the complex number in standard polar form, plot the number in the complex plane, and determine the quadrant in which the point lies or the axis on which the point lies.

a. $z = 3\left(\cos\dfrac{5\pi}{8} + i\sin\dfrac{5\pi}{8}\right)$

b. $z = 2\left(\cos\dfrac{23\pi}{4} + i\sin\dfrac{23\pi}{4}\right)$

c. $z = 4(\cos(-3\pi) + i\sin(-3\pi))$

Section 5.3

## Section 5.3 Objective 3 Converting a Complex Number from Polar Form to Rectangular Form

Work through the video with Example 3 showing all work below.
Write each complex number in rectangular form using exact values if possible. Otherwise, round to two decimal places.

a. $z = 3\left(\cos\dfrac{7\pi}{4} + i\sin\dfrac{7\pi}{4}\right)$

b. $z = 4(\cos 80° + i\sin 80°)$

## Section 5.3 Objective 4 Converting a Complex Number from Rectangular Form to Standard Polar Form

Write down the four cases outlined in **Converting Complex Numbers From Rectangular Form to Standard Polar Form for Complex Numbers Lying Along the Real Axis or Imaginary Axis**.

Section 5.3

Work through the video with Example 4 showing all work below.
Determine the standard polar form of each complex number. Write the argument using radians.

a. $z = 5$

b. $z = -3i$

c. $z = -\sqrt{7}$

d. $z = \dfrac{7}{2}i$

Section 5.3

What are the four steps for **Converting a Complex Number from Rectangular Form to Standard Polar Form for** $a \neq 0$ **and** $b \neq 0$ ?

**Step 1.**

**Step 2.**

**Step 3.**

**Step 4.**

Section 5.3

Work through the interactive video with Example 5 showing all work below.
Determine the standard polar form of each complex number. Write the argument in radians using exact values if possible. Otherwise, round the argument to two decimal places.

a. $z = -2\sqrt{3} + 2i$

b. $z = 4 - 3i$

## Section 5.3 Objective 5 Determining the Product or Quotient of Complex Numbers in Polar Form

How do we compute the **Product and Quotient of Two Complex Numbers Written in Polar Form**?

Work through the video with Example 6 showing all work below.

Let $z_1 = 4\left(\cos\frac{2\pi}{3} + i\sin\frac{2\pi}{3}\right)$ and $z_2 = 5\left(\cos\frac{11\pi}{6} + i\sin\frac{11\pi}{6}\right)$.

Find $z_1 z_2$ and $\dfrac{z_1}{z_2}$ and write the answers in standard polar form.

Section 5.3

## Section 5.3 Objective 6 Using De Moivre's Theorem to Raise a Complex Number to a Power

If $z = r(\cos\theta + i\sin\theta)$, then use the product of two complex numbers formula to find $z^2$.

Now, find $z^3$.

Now, find $z^4$.

**What is De Moivre's Theorem for Finding Powers of a Complex Number?**

Work through the interactive video with Example 7 showing all work below.

a. Find $\left[5\left(\cos\dfrac{3\pi}{4}+i\sin\dfrac{3\pi}{4}\right)\right]^3$ and write your answer in standard polar form.

b. Find $(\sqrt{3}-i)^4$ and write your answer in rectangular form.

Section 5.3

<u>Section 5.3  Objective 7 Using De Moivre's Theorem to Find the Roots of a Complex Number</u>

**What is De Moivre's Theorem for Finding the *n*th Roots of a Complex Number?**

Work through Example 8 showing all work below.
Find the complex cube roots of 8. Write your answers in rectangular form.

If we can find $z_0$, then we can easily determine the remaining $n - 1$ roots by adding _____ to each successive argument.

Work through the video with Example 9 showing all work below.
a. Find the complex fourth roots of $z = 81\left(\cos\dfrac{3\pi}{5} + i\sin\dfrac{3\pi}{5}\right)$.

Write your answers in standard polar form.

Section 5.3

b. Find the complex square roots of $z = -2\sqrt{3} + 2i$.
   Write your answer in rectangular form with each part rounded to two decimal places.

# Section 5.4 Guided Notebook

## 5.4 Vectors
- ☐ Work through Section 5.4 TTK #1–3
- ☐ Work through Section 5.4 Objective 1
- ☐ Work through Section 5.4 Objective 2
- ☐ Work through Section 5.4 Objective 3
- ☐ Work through Section 5.4 Objective 4
- ☐ Work through Section 5.4 Objective 5
- ☐ Work through Section 5.4 Objective 6
- ☐ Work through Section 5.4 Objective 7
- ☐ Work through Section 5.4 Objective 8
- ☐ Work through Section 5.4 Objective 9

## Section 5.4 Vectors

### 5.4 Things To Know

1. Converting a Point from Polar Coordinates to Rectangular Coordinates
Try working through a "You Try It" problem or watch the video.

2. Converting a Point from Rectangular Coordinates to Polar Coordinates
Try working through a "You Try It" problem or watch the interactive video.

3. Converting a Complex Number from Rectangular Form to Standard Polar Form
Try working through a "You Try It" problem or watch the interactive video.

Section 5.4

Section 5.4 Introduction

What is a **vector**?


Section 5.4 Objective 1 Understanding the Geometric Representation of a Vector

What is the **initial point**? What is the **terminal point**?


What is the definition of the **Geometric Representation of a Vector**?


What is the **zero vector**?


What is the definition of the **Magnitude of a Vector Represented Geometrically**?

Work through the video with Example 1 showing all work below.
Determine the magnitude of **v**.

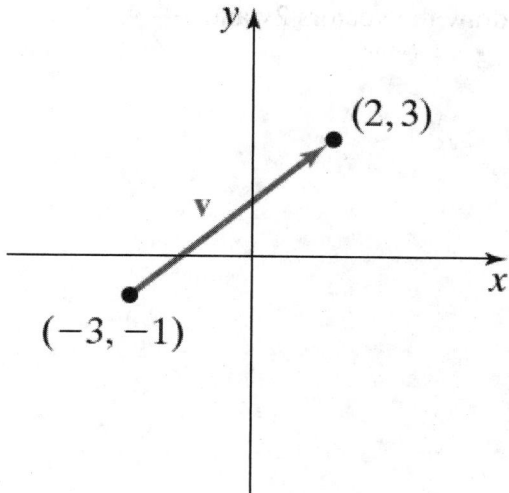

Section 5.4 Objective 2 Understanding Operations on Vector Represented Geometrically

**Scalar Multiplication**
Watch the video associated with scalar multiplication and fill in the blanks below:

The product of a vector *v* and a scalar *k* is the new vector _____.

If $k > 0$, then _____ has the same _____ as **v** with a magnitude of _____.

If $k < 0$, then _____ has the _____ as **v** with a magnitude of _____.

In the video describing scalar multiplication, a vector *v* and *u* are given.
Sketch the vectors $\frac{1}{2}v$ and $-3u$.

Section 5.4

Work through Example 2 showing all work below.

Given the vector **v**, draw the vectors 2**v** and $-\dfrac{1}{2}$**v**.

What is the definition of **Parallel Vectors**?

**Vector Addition**
Watch the video associated with vector addition and draw vector **u** + **v** below.

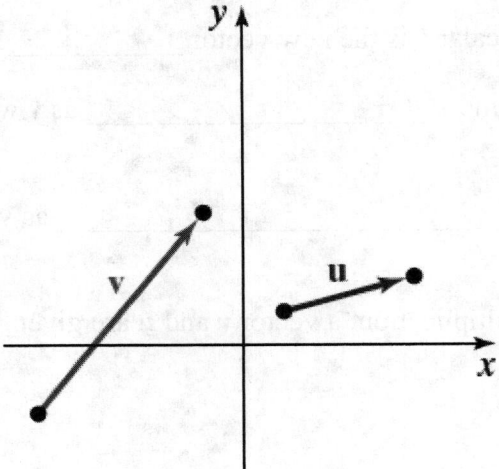

The vector **u** + **v** is called the _____ vector.

In your picture of **u** + **v** above, you may see why vector addition is sometimes called the

_____ law.

Section 5.4

## The **Parallelogram Law for Vector Addition**
Watch the video associated with the parallelogram law for vector addition and show that vector **u + v = v + u**.

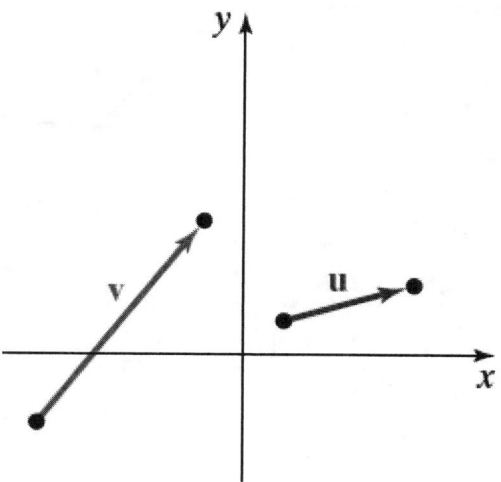

The fact that **u + v = v + u** is known as the _____ property of vector addition.

## Vector Subtraction
Watch the video associated with vector subtraction and draw vector **u − v** below.

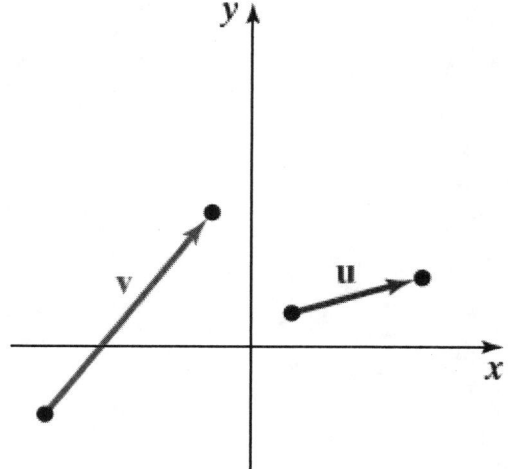

Section 5.4

Work through the video with Example 3 showing all work below.
Given the vectors **u, v,** and **w,** draw each of the following vectors.

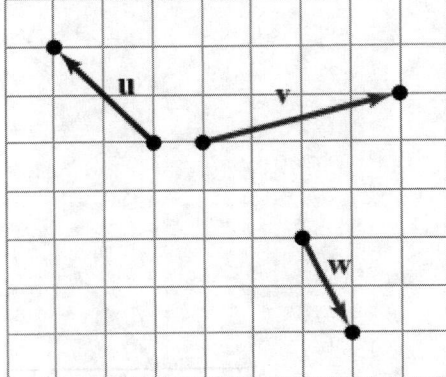

a. **w + v**

b. **v − u**

c. **v + 2w − u**

Section 5.4

## Section 5.4 Objective 3 Understanding Vectors in Terms of Components

What is the definition of **Equal Vectors**?

What is the definition of **Representing a Vector in Terms of Components** *a* **and** *b*?

Work through Example 4 and show all work below.
Determine the component representation and the magnitude of a vector **v** having an initial point $P(5,3)$ and a terminal point $Q(-6,5)$.

What is the definition of a **Vector in Standard Position**?

## Section 5.4  Objective 4 Understanding Vectors Represented in Terms of **i** and **j**

What is the definition of a **Unit Vector**?

What is the definition of the **Unit Vectors i and j**?

What is the definition of a **Vector Represented in Terms of i and j**?

Section 5.4

- Work through Example 5 and show all work below.
Write the vector $\mathbf{v} = \langle -5, 2 \rangle$ in terms of the unit vectors $\mathbf{i}$ and $\mathbf{j}$.

What are the **Operations with Vectors in Terms of i and j**?

- Work through the video with Example 6 and show all work below.
Let $\mathbf{u} = -3\mathbf{i} + 7\mathbf{j}$ and $\mathbf{v} = 5\mathbf{i} - \mathbf{j}$. Find each vector in terms of $\mathbf{i}$ and $\mathbf{j}$ and determine the magnitude of each vector.

a. $-\dfrac{1}{2}\mathbf{u}$

b. $\mathbf{u} + \mathbf{v}$

c. $\mathbf{u} - \mathbf{v}$

Section 5.4

d. 3**u** − 5**v**

**What are the ten Properties of Vectors?**

1.

2.

3.

4.

5.

6.

7.

8.

9.

10.

Section 5.4

## Section 5.4 Objective 5 Finding a Unit Vector

What is the definition of the **Unit Vector in the Same Direction of a Given Vector**?

Work through the video with Example 7 and show all work below.
Find the unit vector that has the same direction as **v = 6i − 8j.**

## Section 5.4 Objective 6 Determining the Direction Angle of a Vector

What is the definition of the **Direction Angle of a Vector**?

Section 5.4

Work through the video with Example 8 and show all work below.
Determine the direction angle of the vector $\mathbf{v} = -3\mathbf{i} + 2\mathbf{j}$.

Section 5.4  Objective 7 Representing a Vector in Terms of **i** and **j** Given Its Magnitude and Direction Angle

Describe the procedure for **Representing a Vector in Terms of i and j Given Its Magnitude and Direction Angle**.

Work through the video with Example 9 and show all work below.
The vector **v** has a magnitude of 20 units and direction angle of $\theta = 50°$.
Represent this vector in the form **v = ai + bj**. Round $a$ and $b$ to two decimal places.

## Section 5.4 Objective 8 Using Vectors to Solve Applications Involving Velocity

Work through the video with Example 10 and show all work below.
An airplane takes off from a runway at a speed of 190 mph at an angle of 11°. Express the velocity of the plane at takeoff as a vector in terms of **i** and **j**. Round $a$ and $b$ to two decimal places.

Section 5.4

Work through the video with Example 11 and show all work below.
The wind is blowing at a speed of 35 mph in a direction of N 30° E. Express the velocity of the wind as a vector in terms of **i** and **j**.

Work through the video with Example 12 and show all work below.
A 747 jet was heading due east 520 mph in still air and encountered a 60 mph headwind blowing in the direction N 40° W. Determine the resulting ground speed of the plane and its new bearing. Round the ground speed to the nearest hundredth.

Section 5.4 Objective 9 Using Vectors to Solve Applications Involving Force

What is the definition of **Static Equilibrium**?

Work through the video with Example 13 and show all work below.
The forces $F_1 = 6i - 8j$ and $F_2 = 3i + 2j$ are acting on an object. What additional force is required for the object to be in static equilibrium?

Work through the video with Example 14 and show all work below.
Two tugboats are towing a large ship out of port and into the open sea. One tugboat exerts a force of $\|F_1\| = 2000$ pounds in a direction N 35° W. The other tugboat pulls with a force of $\|F_2\| = 1400$ pounds in a direction S 55° W. Find the magnitude of the resultant force and the bearing of the ship.

## Section 5.5 Guided Notebook

### 5.5 The Dot Product
- ☐ Work through Section 5.5 TTK #1–4
- ☐ Work through Section 5.5 Objective 1
- ☐ Work through Section 5.5 Objective 2
- ☐ Work through Section 5.5 Objective 3
- ☐ Work through Section 5.5 Objective 4
- ☐ Work through Section 5.5 Objective 5
- ☐ Work through Section 5.5 Objective 6

### Section 5.5 The Dot Product

#### 5.5 Things To Know

1. Understanding Vectors in Terms of **i** and **j**
Try working through a "You Try It" problem or watch the video.




2. Finding a Unit Vector
Try working through a "You Try It" problem or watch the video.




3. Determining the Direction Angle of a Vector
Try working through a "You Try It" problem or watch the video.




4. Representing a Vector in Terms of **i** and **j**
Try working through a "You Try It" problem or watch the video.

Section 5.5

## Section 5.5 Objective 1 Understanding the Dot Product and Its Properties

What is the definition of the **Dot Product of Two Vectors**?

What are two other names for the dot product?

Work through Example 1 showing all work below.
If $\mathbf{u} = -3\mathbf{i} + 5\mathbf{j}$ and $\mathbf{v} = 7\mathbf{i} - 4\mathbf{j}$, then find $\mathbf{u} \cdot \mathbf{v}$ and $\mathbf{v} \cdot \mathbf{u}$.

What are the five **Dot Product Properties**?

1.

2.

3.

4.

5.

Section 5.5

Work through the video with Example 2 showing all work below.
If $\mathbf{u} = -4\mathbf{i} + 6\mathbf{j}$, $\mathbf{v} = -2\mathbf{i} + 8\mathbf{j}$, and $\mathbf{w} = -3\mathbf{i} - \mathbf{j}$, then find each of the following:

a. $\mathbf{u} \cdot \mathbf{v}$

b. $\mathbf{u} \cdot (\mathbf{v} + \mathbf{w})$

c. $\mathbf{u} \cdot (-5\mathbf{v})$

d. $\|\mathbf{w}\|^2$

## Section 5.5 Objective 2 Using the Dot Product to Determine the Angle between Two Vectors

If $\mathbf{u}$ and $\mathbf{v}$ are non-zero vectors and if $\theta$ is the angle between $\mathbf{u}$ and $\mathbf{v}$, then

_____.

Section 5.5

Work through the video with Example 3 showing all work below.
Determine the angle between each pair of vectors. Give the angle in degrees rounded to the nearest hundredth of a degree.

a. $u = i + 4j$, $v = -2i + 5j$

b. $u = 3i - 2j$, $v = 4i + 6j$

What is the definition of the **Alternate Form of the Dot Product of Two Vectors**?

## Section 5.5 Objective 3 Using the Dot Product to Determine If Two Vector Are Orthogonal or Parallel

What is the definition of **Orthogonal Vectors**?

What is the **Test for Orthogonal Vectors**?

Work through Example 4 and show all work below.
Determine the value of $b$ so that the vectors $\mathbf{u} = \mathbf{i} + b\mathbf{j}$ and $\mathbf{v} = 3\mathbf{i} + 10\mathbf{j}$ are orthogonal.

What is the **Test for Parallel Vectors**?

Section 5.5

Work through the video with Example 5 and show all work below.

Determine if $\mathbf{u} = -\frac{1}{2}\mathbf{i} - \mathbf{j}$ and $\mathbf{v} = 2\mathbf{i} + 4\mathbf{j}$ are orthogonal, parallel, or neither.

Section 5.5 Objective 4 Decomposing a Vector into Two Orthogonal Vectors

Below are two non-zero vectors **v** and **w** that share the same initial point $P$. Draw a line segment from the terminal point of **v** to point $Q$. Now draw vector **PQ**. This vector is called the vector _____ of **v** onto **w**.

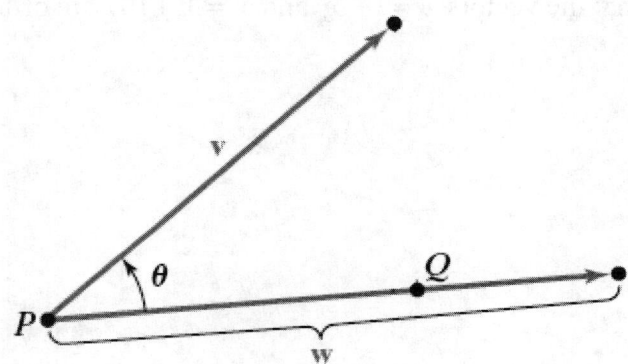

We denote the vector projection of **v** onto **w** as _____.

What is the definition of the **Scalar Component of v in the Direction of w**?

312

Section 5.5

What is the definition of the **Vector Projection of v onto w**?

Work through Example 6 and show all work below.
If $v = i + 2j$ and $w = 4i + j$, determine the vector $\text{proj}_w v$.

What is the definition of the **Vector Decomposition of v into Orthogonal Components**?

Section 5.5

Work through the video with Example 7 and show all work below.
Let $v = i + 2j$ and $w = 4i + j$. Determine the vector decomposition of $v$ into orthogonal components $v_1$ and $v_2$, where $v_1$ is parallel to $w$ and $v_2$ is orthogonal to $w$.

## Section 5.5 Objective 5 Solving Applications Involving Forces on an Inclined Plane

Work through the video with Example 8 and show all work below.
A 200-pound object is placed on a ramp that is inclined at 22°. What is the magnitude of the force needed to hold the box in a stationary position to prevent the box from sliding down the ramp? What is the magnitude of the force pushing against the ramp?

## Section 5.5 Objective 6 Solving Applications Involving Work

What is the definition of **work**?

Work through Example 9 and show all work below.
A horse is pulling a plow with a force of 400 pounds. The angle between the harness and the ground is 20°. How much work is done to pull the plow 50 feet?